Everyday Mathematics®

The University of Chicago School Mathematics Project

Student Math Journal
Volume 1

Grade  5

Wright Group

The University of Chicago School Mathematics Project (UCSMP)

Max Bell, Director, UCSMP Elementary Materials Component; Director, *Everyday Mathematics* First Edition; James McBride, Director, *Everyday Mathematics* Second Edition; Andy Isaacs, Director, *Everyday Mathematics* Third Edition; Amy Dillard, Associate Director, *Everyday Mathematics* Third Edition

Authors

Max Bell, John Bretzlauf, Amy Dillard, Robert Hartfield, Andy Isaacs, James McBride, Kathleen Pitvorec, Peter Saecker, Noreen Winningham*, Robert Balfanz†, William Carroll†

*Third Edition only   †First Edition only

Technical Art	Editorial Assistant	Teachers in Residence
Diana Barrie	Rosina Busse	Fran Goldenberg, Sandra Vitantonio

Photo Credits

©W. Perry Conway/CORBIS, cover, *right*; Getty Images, cover, *bottom left*; ©Andrew Kolb/Masterfile, p. viii; ©PIER/Getty Images, cover, *center*.

Contributors

Tammy Belgrade, Diana Carry, Debra Dawson, Kevin Dorken, James Flanders, Laurel Hallman, Ann Hemwall, Elizabeth Homewood, Linda Klaric, Lee Kornhauser, Judy Korshak-Samuels, Deborah Arron Leslie, Joseph C. Liptak, Sharon McHugh, Janet M. Meyers, Susan Mieli, Donna Nowatzki, Mary O'Boyle, Julie Olson, William D. Pattison, Denise Porter, Loretta Rice, Diana Rivas, Michelle Schiminsky, Sheila Sconiers, Kevin J. Smith, Theresa Sparlin, Laura Sunseri, Kim Van Haitsma, John Wilson, Mary Wilson, Carl Zmola, Theresa Zmola

 This material is based upon work supported by the National Science Foundation under Grant No. ESI-9252984. Any opinions, findings, conclusions, or recommendations expressed in this material are those of the authors and do not necessarily reflect the views of the National Science Foundation.

www.WrightGroup.com

Send all inquiries to:
Wright Group/McGraw-Hill
P.O. Box 812960
Chicago, IL 60681

ISBN 0-07-604603-6

13 14 15 16 17 QWD 14 1312 11 10 09

The McGraw-Hill Companies

Contents

UNIT 1 Number Theory

UNIT 2 Estimation and Computation

UNIT 3 | **Geometry Explorations and the American Tour**

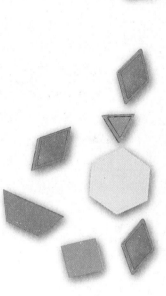

UNIT 4 Division

UNIT 5 Fractions, Decimals, and Percents

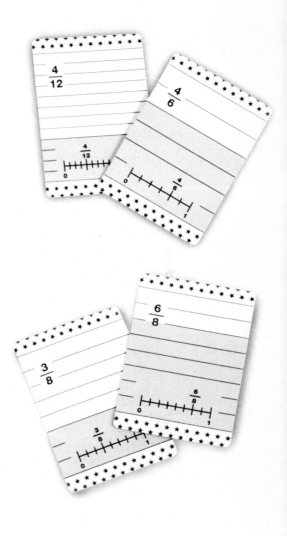

UNIT 6 Using Data; Addition and Subtraction of Fractions

LESSON 1·1 Welcome to *Fifth Grade Everyday Mathematics*

Much of what you have learned up to now in *Everyday Mathematics* has been basic training in mathematics and its uses. This year, you will extend the skills and ideas you have learned, and you will also study other ideas in mathematics—many of which your older brothers and sisters, or even your parents, may not have learned until high school. The authors of *Everyday Mathematics* believe that today's fifth graders can learn more and do more than fifth graders in the past.

Here are some of the things you will be asked to do in *Fifth Grade Everyday Mathematics:*

◆ Practice and extend your knowledge of numbers and their properties, as well as your ability to use measurements and estimation.

◆ Review and extend your skills in doing arithmetic, using a calculator, and thinking about problems and their solutions. You will work with and learn the notations for fractions, decimals, percents, large whole numbers, exponents, and negative numbers.

◆ Continue your work with algebra, using variables in place of numbers to represent and analyze situations.

◆ Refine your understanding of geometry. You will define and classify geometric figures more completely than before. You will construct figures and transformations. You will find the perimeter and area of 2-dimensional shapes, and the volume and surface area of 3-dimensional figures.

◆ You will study the history, people, and environment of the United States through numerical data. You will learn to interpret many kinds of maps, graphs, and tables and use them to solve problems. Look at journal page 2. Without telling anyone, write a secret number in the margin at the top of the page in the right hand corner.

◆ You will use data that comes from questionnaires and experiments to explore probability and statistics.

We want you to become better at using mathematics so you may better understand your world. We hope that you enjoy the activities in *Fifth Grade Everyday Mathematics* and that they will help you appreciate the beauty and usefulness of mathematics in your daily activities.

LESSON 1·1 · *Student Reference Book* Scavenger Hunt

The object of this scavenger hunt is to score as close to 90 total points as you can. Solve the problems on this page and page 3. Use your *Student Reference Book* to find information about each problem, and record the page numbers.

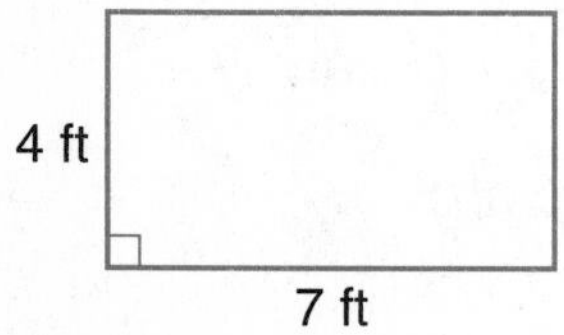

	Problem Points	Page Points

1. Circle the prime numbers in the following list: _____ _____

   1   2   6   9   13   20   31   63   72

   SRB page _____

2. 5 meters = _____ centimeters   _____ _____

   SRB page _____

3. 300 mm = _____ cm   _____ _____

   SRB page _____

4. What is the perimeter of this figure? _____ ft   _____ _____

   4 ft ⌐_____⌐
   7 ft

   SRB page _____

5. Name two fractions equivalent to $\frac{4}{6}$.   _____ _____

   _____ and _____

   SRB page _____ _____ _____

6. Is angle *RST* acute or obtuse? _____   _____ _____

   SRB page _____

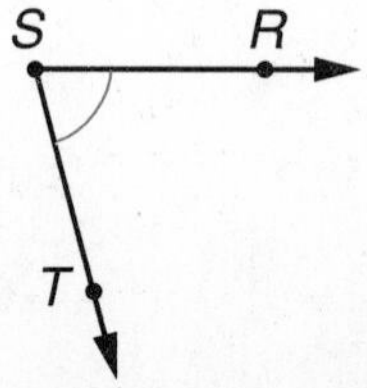

LESSON 1·1

Student Reference Book **Scavenger Hunt** *continued*

	Problem Points	Page Points

7. a. What is the definition of a scalene triangle?   _____   _____

_____

_____

b. Draw and label a scalene triangle.

SRB

page _____

8. What materials do you need to play *Top-It* games?   _____   _____

_____

_____

Choose one of the versions of *Top-It,* and play it with a partner.

SRB

page _____

Record your scavenger hunt scores in the table below. Then calculate the totals.

Problem Number	Problem Points	Page Points	Total Points = Problem Points + Page Points
1			
2			
3			
4			
5			
6			
7			
8			
Total Points			

LESSON 1·1   Math Boxes

1. Next to each *Student Reference Book* icon in Problems 2–6, write the SRB page numbers on which you can find information about each problem.

2. Write a 6-digit numeral that has
4 in the hundredths place,
5 in the thousands place,
4 in the hundreds place,
and 0 in all other places.

—, — — —. — —

3. List all the factors of 12.

_____

4. Write the multiplication/division fact family for 2, 7, and 14.

_____

_____

_____

_____

5. Write a 7-digit numeral that has
7 in the ones place,
8 in the millions place,
4 in the ten-thousands place,
and 0 in all other places.

—, — — —, — — —

6. Solve.

a.   127
         250
       + 63

b.   105
        − 59

LESSON 1·2 Arrays

A **rectangular array** is an arrangement of objects into rows and columns. Each row has the same number of objects, and each column has the same number of objects.

A multiplication **number model** can be written to describe a rectangular array. The first factor is the number of rows in the array. The second factor is the number of columns. The product is the total number of objects.

This is an array of 8 dots.
It has 4 rows with 2 dots in each row.
It has 2 columns with 4 dots in each column.

The number model is next to the array.

$4 * 2 = 8$

This is another array of 8 dots.
It has 2 rows with 4 dots in each row.
It has 4 columns with 2 dots in each column.

Label this array by writing the number model next to it.

1. a. Take 10 counters. Make as many different rectangular arrays as you can using all 10 counters.

   b. Draw each array on the grid at the right by marking dots.

   c. Write the number model next to each array.

2. a. How many dots are in the array at the right?

   _____

   b. Write a number model for the array.

   _____

   c. Make as many other arrays as you can with the same number of dots that were used for the array in Part 2a. Draw each array on the grid at the right. Write a number model for each array.

LESSON 1·2 Addition and Subtraction Fact Families

1. Find and read the information about **Fact Family** in the glossary of the *Student Reference Book.* There are four related facts in a fact family. In an addition and subtraction fact family, there are two addition facts and two subtraction facts.

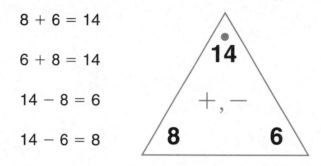

$8 + 6 = 14$

$6 + 8 = 14$

$14 - 8 = 6$

$14 - 6 = 8$

Knowing how the facts in fact families are related can help you solve problems.

If you know two of the numbers in a subtraction problem, you can solve one of the addition facts to find the missing number.

$14 - 8 = \square$ *Think:* $8 + \square = 14$

2. Write the fact family represented by each of these fact triangles.

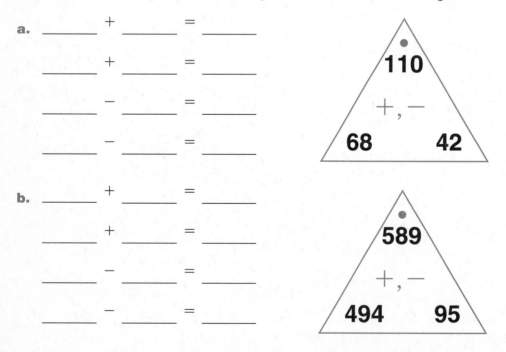

a. _____ + _____ = _____

    _____ + _____ = _____

    _____ − _____ = _____

    _____ − _____ = _____

b. _____ + _____ = _____

    _____ + _____ = _____

    _____ − _____ = _____

    _____ − _____ = _____

Addition and Subtraction Fact Families *cont.*

c. _____ + _____ = _____

_____ + _____ = _____

_____ − _____ = _____

_____ − _____ = _____

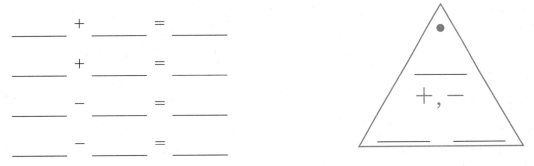

1,647
+, −
984   663

Solve using the fact families from Problem 2.

3. a. $110 - $ _____ $= 42$

b. $589 - $ _____ $= 494$

4. Complete your own fact triangle and write the fact family that it represents.

_____ + _____ = _____

_____ + _____ = _____

_____ − _____ = _____

_____ − _____ = _____

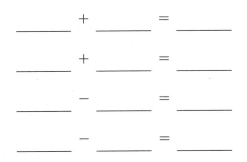

_____
+, −
_____   _____

5. Three of the following numbers can be used in an addition and subtraction fact family. Find the numbers and write the fact family.

56, 29, 78, 212, 134, 2

_____ + _____ = _____

_____ + _____ = _____

_____ − _____ = _____

_____ − _____ = _____

LESSON 1·2 Math Boxes

1. Marcus drew 8 cards from a pile: 10, 8, 4, 5, 8, 6, 12, and 1. Find the following landmarks:

a. Maximum _____

b. Minimum _____

c. Range _____

d. Median _____

SRB 119

2. Name five numbers between 0 and 1.

_____

_____

_____

_____

_____

SRB 26 56

3. Make an array for each of these number sentences.

a. 3 * 9 = 27

b. 6 * 7 = 42

SRB 10

4. a. Write the largest number you can make using each of the digits 7, 1, 0, 2, and 9 just once.

_____

b. Write the smallest number. (Do not start with 0.)

_____

SRB 4

5. Draw a line from each spinner to the number that represents the shaded parts.

$\dfrac{1}{3}$ $\dfrac{1}{4}$ 0.75   50%

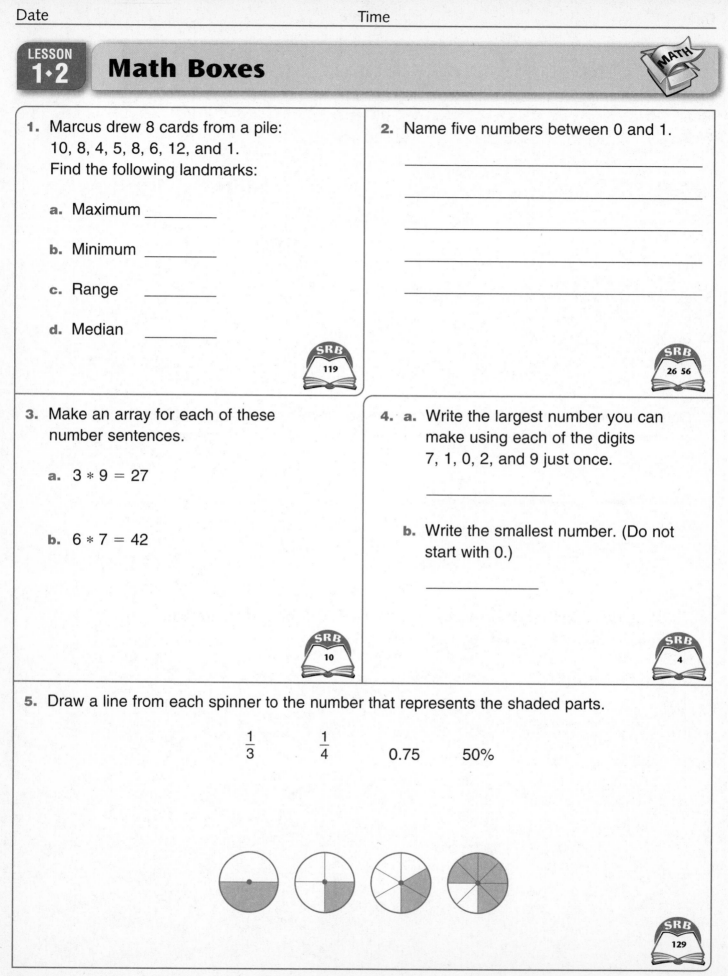

SRB 129

8

LESSON 1·3 Multiplication Facts Master Lists

Make a check mark next to each fact you missed and need to study.
When you have learned a fact, write "OK" next to the check mark.

Multiplication Facts			
3s	5s	7s	9s
3 * 3	5 * 3	7 * 3	9 * 3
3 * 4	5 * 4	7 * 4	9 * 4
3 * 5	5 * 5	7 * 5	9 * 5
3 * 6	5 * 6	7 * 6	9 * 6
3 * 7	5 * 7	7 * 7	9 * 7
3 * 8	5 * 8	7 * 8	9 * 8
3 * 9	5 * 9	7 * 9	9 * 9
		7 * 10	9 * 10

Multiplication Facts			
4s	6s	8s	10s
4 * 3	6 * 3	8 * 3	10 * 3
4 * 4	6 * 4	8 * 4	10 * 4
4 * 5	6 * 5	8 * 5	10 * 5
4 * 6	6 * 6	8 * 6	10 * 6
4 * 7	6 * 7	8 * 7	10 * 7
4 * 8	6 * 8	8 * 8	10 * 8
4 * 9	6 * 9	8 * 9	10 * 9
	6 * 10	8 * 10	10 * 10

LESSON 1·3 Factor Pairs

Math Message

A 2-row-by-5-column array

$$2 * 5 = 10$$

Factors   Product

$2 * 5 = 10$ is a number model for the 2-by-5 array.

10 is the **product** of 2 and 5.

2 and 5 are whole-number **factors** of 10.

2 and 5 are a **factor pair** for 10.

1 and 10 are also factors of 10 because $1 * 10 = 10$.

1 and 10 are another **factor pair** for 10.

1. **a.** Use counters to make all possible arrays for the number 14.

   b. Write a number model for each array you make.

   _____

   _____

   c. List all the whole-number factors of 14.

   _____

2. Write number models to help you find all the factors of each number below.

Number	Number Models with 2 Factors	All Possible Factors
20		
16		
13		
27		
32		

LESSON 1·3 Math Boxes

1. Where in the *Student Reference Book* would you look to find the definition of *factor pair?*
Fill in the circle next to the best answer.

- (A) Table of Contents
- (B) Index
- (C) Glossary
- (D) Whole Numbers Section

SRB 10

2. Write a 6-digit numeral with
4 in the hundreds place,
8 in the hundred-thousands place,
3 in the ones place,
and 7s in all other places.

—— —— ——,—— —— ——

SRB 28

3. List all the factors of 20.

_____

SRB 12

4.a. Complete the fact triangle.

b. Write the fact family for this triangle.

30

*, /

5 _____

_____ = _____

_____ = _____

_____ = _____

_____ = _____

SRB 219 412

5. Write a 7-digit numeral with

6 in the ones place,
3 in the thousandths place,
1 in the thousands place,
2 in the tenths place,
and 0s in all other places.

____,____ ____ ____.____ ____ ____

SRB 4 27–31

6. Add or subtract.

a. $67 + 109 + 318 =$ _____

b. $2,005 - 189 =$ _____

c. $39 + 71 + 177 =$ _____

d. $40,031 - 277 =$ _____

SRB 13–17

LESSON 1·4   Math Boxes

1. Find the following landmarks for the set of numbers: 28, 17, 45, 32, 29, 28, 14, 27.

   a. Maximum _____

   b. Minimum _____

   c. Range _____

   d. Median _____

   SRB 119

2. Write five positive numbers that are less than 2.5.

   _____

   _____

   _____

   _____

   _____

   SRB 26

3. a. Make an array for the number sentence 4 * 8 = 32.

   b. Write a number story for the number sentence.

   _____

   _____

   _____

   SRB 6 10

4. a. What is the smallest whole number you can make using each of the digits 5, 8, 2, 7, and 4 just once?

   _____

   b. What is the largest?

   _____

   SRB 4

5. Draw a line from each spinner to the number that represents the shaded parts.

   $66\frac{2}{3}\%$ $\frac{1}{2}$ 0.625 $\frac{2}{8}$

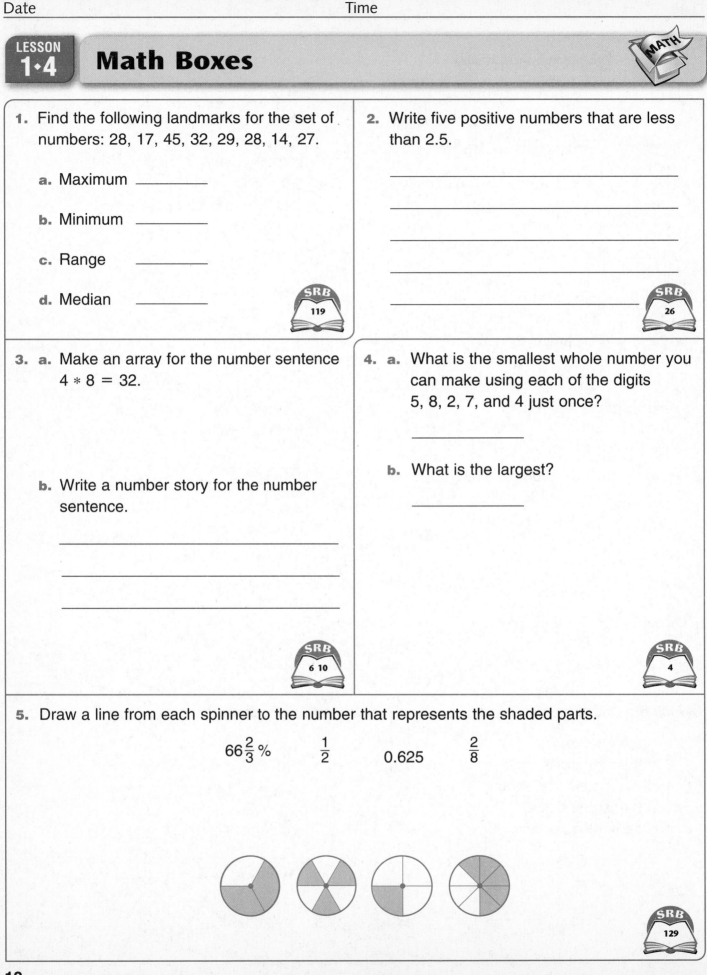

   SRB 129

Divisibility

Math Message

1. Circle the numbers that are divisible by 2.

   28   57   33   112   123,456   211   5,374   900   399   705

2. What do the numbers that you circled have in common?

   _____

   _____

   _____

Suppose you divide a whole number by a second whole number. The answer may be a whole number, or it may be a number that has a decimal part. If the answer is a whole number, we say that the first number is **divisible by** the second number. If the answer has a decimal part, the first number is *not* divisible by the second number.

Example 1: *Is 135 divisible by 5?*
To find out, divide 135 by 5.

$$135 / 5 = 27$$

The answer, 27, is a whole number. So 135 is divisible by 5.

Example 2: *Is 122 divisible by 5?*
To find out, divide 122 by 5.

$$122 / 5 = 24.4$$

The answer, 24.4, has a decimal part. So 122 is *not* divisible by 5.

Use your calculator to help you answer these questions.

3. Is 267 divisible by 9?   _____

4. Is 552 divisible by 6?   _____

5. Is 809 divisible by 7?   _____

6. Is 7,002 divisible by 3?   _____

7. Is 4,735 divisible by 5?   _____

8. Is 21,733 divisible by 4?   _____

9. Is 5,268 divisible by 22?   _____

10. Is 2,072 divisible by 37?   _____

LESSON 1·5 — Divisibility Rules

For many numbers, even large ones, it is possible to test for divisibility without actually dividing.

Here are the most useful divisibility rules:

◆ All numbers are **divisible by 1.**

◆ All even numbers (ending in 0, 2, 4, 6, or 8) are **divisible by 2.**

◆ A number is **divisible by 3** if the sum of its digits is divisible by 3.
   Example: 246 is divisible by 3 because 2 + 4 + 6 = 12, and 12 is divisible by 3.

◆ A number is **divisible by 6** if it is divisible by both 2 and 3.
   Example: 246 is divisible by 6 because it is divisible by 2 and by 3.

◆ A number is **divisible by 9** if the sum of its digits is divisible by 9.
   Example: 51,372 is divisible by 9 because 5 + 1 + 3 + 7 + 2 = 18, and
             18 is divisible by 9.

◆ A number is **divisible by 5** if it ends in 0 or 5.

◆ A number is **divisible by 10** if it ends in 0.

1. Test each number below for divisibility. Then check on your calculator.

Number	Divisible... by 2?	by 3?	by 6?	by 9?	by 5?	by 10?
75		✓			✓	
7,960						
384						
3,725						
90						
36,297						

2. Find a 3-digit number that is divisible by both 3 and 5.

_____

3. Find a 4-digit number that is divisible by both 6 and 9.

_____

LESSON 1·5 Math Boxes

1. Circle the numbers that are divisible by 3.

221   381   474   922   726

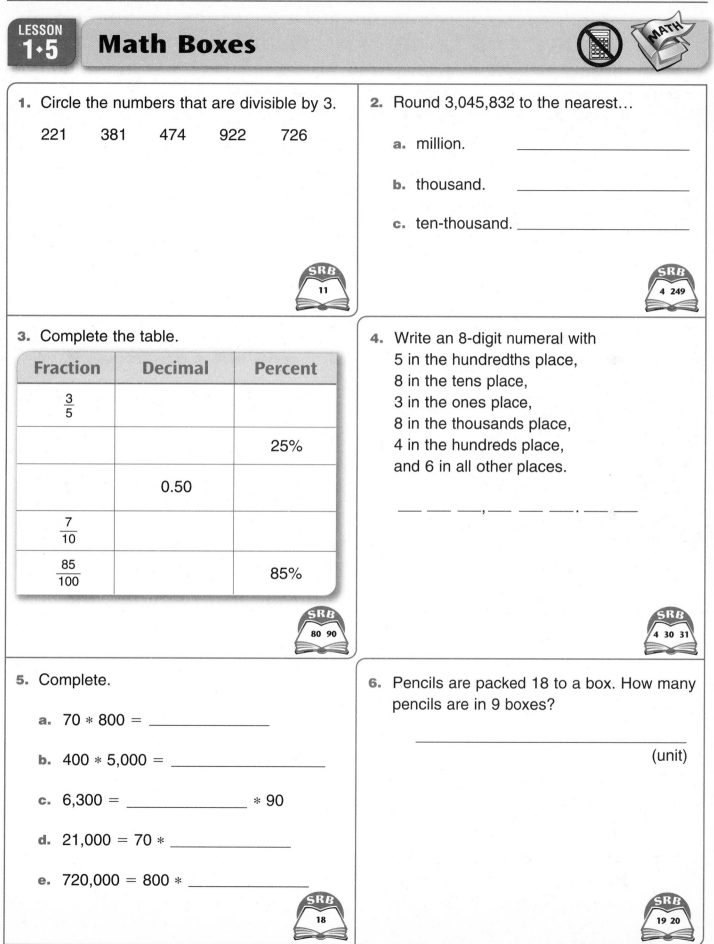

SRB
11

2. Round 3,045,832 to the nearest…

a. million.   _____

b. thousand.   _____

c. ten-thousand. _____

SRB
4 249

3. Complete the table.

Fraction	Decimal	Percent
$\frac{3}{5}$		
		25%
	0.50	
$\frac{7}{10}$		
$\frac{85}{100}$		85%

SRB
80 90

4. Write an 8-digit numeral with
5 in the hundredths place,
8 in the tens place,
3 in the ones place,
8 in the thousands place,
4 in the hundreds place,
and 6 in all other places.

___ ___ ___,___ ___ ___ . ___ ___

SRB
4 30 31

5. Complete.

a. 70 * 800 = _____

b. 400 * 5,000 = _____

c. 6,300 = _____ * 90

d. 21,000 = 70 * _____

e. 720,000 = 800 * _____

SRB
18

6. Pencils are packed 18 to a box. How many pencils are in 9 boxes?

_____
(unit)

SRB
19 20

15

LESSON 1·6 Prime and Composite Numbers

A **prime number** has exactly two factors—1 and the number itself.
A **composite number** has more than two factors.

1. List all the factors of each number in the table. Write *P* if it is a prime number or *C* if it is a composite number.

Number	Factors	P or C	Number	Factors	P or C
2			21		
3	1, 3	P	22		
4			23		
5			24		
6	1, 2, 3, 6	C	25	1, 5, 25	C
7			26		
8			27		
9			28		
10			29		
11			30		
12			31		
13			32		
14			33		
15			34		
16			35		
17			36		
18			37		
19	1, 19	P	38		
20			39		

2. How many factors does each prime number have? _____

3. Can a composite number have exactly 2 factors? _____

   If yes, give an example of such a composite number. _____

_____

LESSON 1·6 *Factor Captor* **Strategies**

Work alone to answer the questions below. Then compare your answers to your partner's. If your answers don't agree with your partner's answers, try to convince your partner that your answers are correct.

1	2	3	4	5	6	7	8	9	10
11	12	13	14	15	16	17	18	19	20
21	22	23	24	25	26	27	28	29	30

1. Suppose you played *Factor Captor* using the above number grid. No numbers have been covered yet. Which is the best number choice you could make? Why?

_____

_____

_____

2. Suppose the 29 and 1 squares have already been covered. Which is the best number choice you could make? Why?

_____

_____

_____

3. Suppose that the 29, 23, and 1 squares have already been covered. Which is the best number choice you could make? Why?

_____

_____

_____

LESSON 1·6 Number-Line Patterns

Find the patterns and fill in the missing values on the number lines.

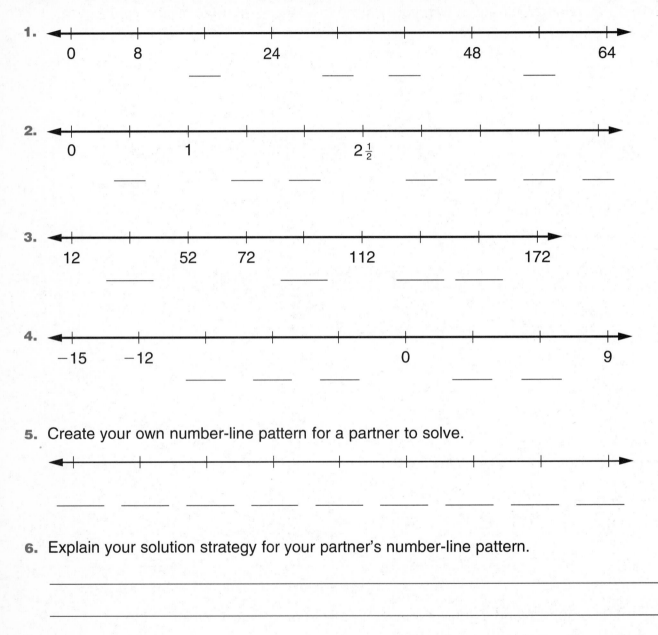

1. 0   8   24   48   64

2. 0   1 $2\frac{1}{2}$

3. 12   52   72   112   172

4. -15 -12 0   9

5. Create your own number-line pattern for a partner to solve.

6. Explain your solution strategy for your partner's number-line pattern.

_____

_____

_____

_____

_____

LESSON 1·6
Math Boxes

1. Write < or >.

   a. 0.5 _____ 1.0

   b. 3.2 _____ 3.02

   c. 4.83 _____ 4.8

   d. 6.25 _____ 6.4

   e. 0.7 _____ 0.07

SRB 9 32 33

2. Round each number to the nearest ten-thousand.

   a. 92,856 _____

   b. 108,325 _____

   c. 5,087,739 _____

   d. 986,402 _____

   e. 397,506 _____

SRB 4 249

3. List all the factors of 36.

   _____

   _____

SRB 10 12

4. Math class ends at 2:20 P.M. It is 1:53 P.M. How many more minutes before math class ends?

   _____ (unit)

SRB 244 245

5. Subtract. Show your work.

SRB 15–17

   a. 105 − 59 = _____

   b. 2,005 − 189 = _____

   c. 680 − 74 = _____

   d. 3,138 − 809 = _____

LESSON 1·7   Square Numbers

A **square array** is a special rectangular array that has the same number of rows as it has columns. A square array represents a whole number, called a **square number.**

The first four square numbers and their arrays are shown below.

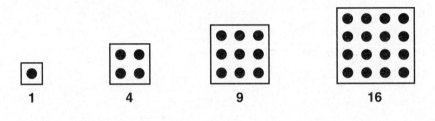

1   4   9   16

1. Draw a square array for the next square number after 16.

Square number: _____

2. List all the square numbers through 100. Use counters or draw arrays if you need help.

_____

3. Can a square number be a prime number? _____ Why or why not?

_____

_____

4. Notice which square numbers are even and which ones are odd.

Can you find a pattern? _____

If yes, describe the pattern.

_____

_____

LESSON 1·7	**Square Numbers** *continued*

Any square number can be written as the product of a number multiplied by itself.

Example: The third square number, 9, can be written as 3 * 3.

There is a shorthand way of writing square numbers: $9 = 3 * 3 = 3^2$.

You can read 3^2 as "3 times 3," "3 squared," or "3 to the second power." The raised 2 is called an **exponent.** It tells that 3 is used as a factor 2 times. Numbers written with an exponent are said to be in **exponential notation.**

Be careful! The number 3^2 is not the same as the product of 3 * 2. $3 * 2 = 6$, but 3^2 equals 3 * 3, which is 9.

5. Fill in the blanks.

Product	Exponential Notation	Square Number
4 * 4	4^2	16
7 * 7		
10 * 10		
____ * ____	11^2	

Some calculators have a key that can be used to find the square of a number. Depending on the calculator, this key might be marked with the symbol ⌃ or the symbol x^2.

6. Use your calculator to find 3 squared.
Write the key sequence that you used. _____

What does the display show? _____

If your calculator has an exponent key, use it to solve the following problems. If not, you can use the multiplication key.

7. $8^2 = $ _____ **8.** $12^2 = $ _____ **9.** $14^2 = $ _____

10. $20^2 = $ _____ **11.** $43^2 = $ _____ **12.** $67^2 = $ _____

13. Start with 4. Square it. Now square the result. What do you get? _____

Math Boxes

1. Circle the numbers that are divisible by 6.

    438   629   702   320   843

SRB
11

2. Round 15,783,406 to the nearest…

    a. million.   _____

    b. thousand.   _____

    c. hundred-thousand. _____

SRB
4 249

3. Complete the table.

Fraction	Decimal	Percent
$\frac{1}{2}$		
	0.125	
	0.80	
$\frac{3}{4}$		
		32%

SRB
80 90

4. **a.** Write a 6-digit numeral with
4 in the hundredths place,
3 in the hundreds place,
6 in the thousands place,
5 in the tens place,
and 2s in all other places.

    ___ , ___ ___ ___ . ___ ___

    b. Write this numeral in words.

    _____

    _____

    _____

SRB
4 30 31

5. Complete.

    a. $900 * 800 =$ _____

    b. $5,000 *$ _____ $= 300,000$

    c. $5,400 =$ _____ $* 60$

    d. $42,000 =$ _____ $* 700$

    e. $1,500 =$ _____ $* 3$

SRB
18

6. **a.** How many crayons are in 10 boxes
if each box contains 48 crayons?

    _____ _____
                                 (unit)

    b. How many crayons would be in
1,000 boxes?

    _____ _____
                                 (unit)

SRB
18–20

LESSON 1·8

Unsquaring Numbers

You know that $6^2 = 6 * 6 = 36$. The number 36 is called the **square** of 6. If you **unsquare** 36, the result is 6. The number 6 is called the **square root** of 36.

1. Unsquare each number. The result is its square root. Do not use the ⬜√ key on your calculator.

   Example: $\underline{12}^2 = 144$ The square root of 144 is $\underline{12}$.

   a. $\underline{}^2 = 225$ The square root of 225 is $\underline{}$.

   b. $\underline{}^2 = 729$ The square root of 729 is $\underline{}$.

   c. $\underline{}^2 = 1{,}600$ The square root of 1,600 is $\underline{}$.

   d. $\underline{}^2 = 361$ The square root of 361 is $\underline{}$.

2. Which of the following are square numbers? Circle them.

   576   794   1,044   4,356   6,400   5,770

List all factors of each square number. Make a factor rainbow to check your work. Then fill in the missing numbers.

3. 49: $\underline{}^2 = 49$ The square root of 49 is $\underline{}$.

4. 64: $\underline{}^2 = 64$ The square root of 64 is $\underline{}$.

5. 81: $\underline{}^2 = 81$ The square root of 81 is $\underline{}$.

6. 100: $\underline{}^2 = 100$

   The square root of 100 is $\underline{}$.

LESSON 1·8

Math Boxes

1. Write < or >.

a. 3.8 _____ 0.83

b. 0.4 _____ 0.30

c. 6.24 _____ 6.08

d. 0.05 _____ 0.5

e. 7.12 _____ 7.2

SRB
9 32 33

2. Round each number to the nearest thousand.

a. 8,692 _____

b. 49,573 _____

c. 2,601,458 _____

d. 300,297 _____

e. 599,999 _____

SRB
4 249

3. List all the factors of 64.

_____

SRB
10 12

4. In the morning, I need 30 minutes to shower and dress, 15 minutes to eat, and another 15 minutes to ride my bike to school. School begins at 8:30 A.M. What is the latest time I can get up and still get to school on time?

_____

SRB
244 245

5. Subtract. Show your work.

a. 777
   − 259

b. 555
   − 125

c. 5,009
   − 188

d. 8,435
   − 997

SRB
15–17

LESSON 1·9 Factor Strings

A **factor string** is a name for a number written as a product of two or more factors. In a factor string, 1 may not be used as a factor.

The **length of a factor string** is equal to the number of factors in the string. The longest factor string for a number is made up of prime numbers. The longest factor string for a number is called the **prime factorization** of that number.

Example:

Number	Factor Strings	Length
20	2 * 10	2
	4 * 5	2
	2 * 2 * 5	3

The order of the factors is not important.
For example, 2 * 10 and 10 * 2 are the same factor string.

The longest factor string for 20 is 2 * 2 * 5.
So the prime factorization of 20 is 2 * 2 * 5.

1. Find all the factor strings for each number below.

a.

Number	Factor Strings	Length
12		

b.

Number	Factor Strings	Length
16		

c.

Number	Factor Strings	Length
18		

d.

Number	Factor Strings	Length
28		

LESSON 1·9   **Factor Strings** *continued*

2. Write the prime factorization (the longest factor string) for each number.

   a. 27 = _____

   b. 40 = _____

   c. 36 = _____

   d. 42 = _____

   e. 48 = _____

   f. 60 = _____

   g. 100 = _____

An **exponent** is a raised number that shows how many times the number
to its left is used as a factor.

Examples: $5^2 \leftarrow$ exponent 5^2 means 5 * 5, which is 25.

   5^2 is read as "5 squared" or as "5 to the second power."

   $10^3 \leftarrow$ exponent 10^3 means 10 * 10 * 10, which is 1,000.

   10^3 is read as "10 cubed" or as "10 to the third power."

   $2^4 \leftarrow$ exponent 2^4 means 2 * 2 * 2 * 2, which is 16.

   2^4 is read as "2 to the fourth power."

3. Rewrite each number written in exponential notation as a product of factors.
   Then find the answer.

   Examples: 2^3 = _____ *2 * 2 * 2* _____ = _____ *8* _____

   $2^2 * 9$ = _____ *2 * 2 * 9* _____ = _____ *36* _____

   a. 10^4 = _____ = _____

   b. $3^2 * 5$ = _____ = _____

   c. $2^4 * 10^2$ = _____ = _____

4. Rewrite each product using exponents.

   Examples: 5 * 5 * 5 = _____ *5^3* _____   5 * 5 * 3 * 3 = _____ *$5^2 * 3^2$* _____

   a. 3 * 3 * 3 * 3 = _____

   b. 4 * 7 * 7 = _____

   c. 2 * 5 * 5 * 7 = _____

   d. 2 * 2 * 2 * 5 * 5 = _____

LESSON 1·9 Math Boxes

1. Circle the numbers that are divisible by 9.

360   252   819   426   651

SRB 11

2. Round 385.27 to the nearest …

a. hundred. _____

b. whole number. _____

c. tenth. _____

SRB 4 28–30 249

3. Complete the table.

Fraction	Decimal	Percent
$\frac{3}{8}$		
		60%
$\frac{2}{5}$		
	0.55	
$\frac{8}{100}$		

SRB 80 90

4. a. Write a 6-digit numeral with
7 in the thousands place,
5 in the hundredths place,
4 in the tenths place,
3 in the tens place,
and 9s in all other places.

___, ___ ___ ___ . ___ ___

b. Write this numeral in words.

_____

_____

_____

SRB 4 30 31

5. Complete.

a. 300 * 40 = _____

b. _____ = 80 * 200

c. _____ = 900 * 600

d. 6,400 = _____ * 80

e. 36,000 = 600 * _____

SRB 18

6. a. How many marbles are in 7 bags
if each bag contains 8 marbles?

_____
(unit)

b. How many marbles are in 700 bags
if each bag contains 8 marbles?

_____
(unit)

SRB 18–20

Math Boxes

1. **a.** Write a 7-digit numeral with
   3 in the tens place,
   5 in the hundredths place,
   7 in the hundreds place,
   2 in the ten-thousands place,
   and 4s in all other places.

   ___ ___, ___ ___ ___ . ___ ___

   b. Write this numeral in words.

   _____
   _____
   _____

   SRB 4 28–30

2. Phoebe received these math test scores:
   93, 96, 85, 100, 98, 100, 99, 95.
   Find the following landmarks:

   a. Maximum _____

   b. Minimum _____

   c. Range _____

   d. Median _____

   SRB 119

3. Complete.

   a. $27,000 =$ _____ $* 90$

   b. _____ $= 800 * 600$

   c. _____ $= 700 * 8,000$

   d. _____ $= 50 * 600$

   e. $350 = 7 *$ _____

   SRB 219

4. Write < or >.

   a. 0.90 _____ 0.89

   b. 3.52 _____ 3.8

   c. 6.91 _____ 6.3

   d. 4.05 _____ 4.2

   e. 0.38 _____ 0.5

   SRB 9 32 33

5. Solve.

   a.   207
          − 158

   b.   325
          + 116

   c.   829
          + 580

   d.   628
          − 444

   e.   385
          − 179

   f.   523
          + 478

   SRB 13–17

LESSON 2·1   Estimation Challenge

Sometimes you will be asked to solve a problem for which it is difficult or even impossible to find an exact answer. Your job will be to make your best estimate and then defend it. We call this kind of problem an Estimation Challenge.

Estimation challenges can be difficult, and they take time to solve. Usually, you will work with a partner or as part of a small group.

Estimation Challenge Problem

Imagine that you are living in a time when there are no cars, trains, or planes. You do not own a horse, a boat, or any other means of transportation.

You plan to travel to _____ . You will have to walk there.
(location given by your teacher)

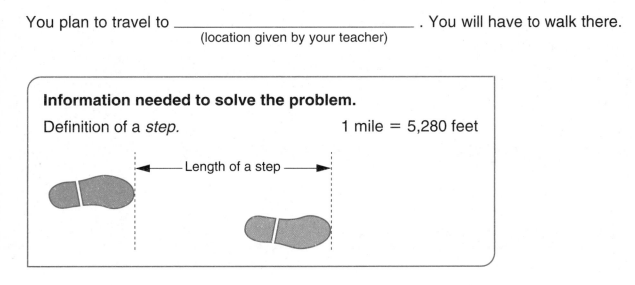

Information needed to solve the problem.

Definition of a *step.*   1 mile = 5,280 feet

Length of a step

1. About how many miles is it from your school to your destination?

   About _____ miles

2. a. About how many footsteps will you have to take to get from your school to your destination?

   About _____ footsteps

   b. What did you do to estimate the number of footsteps you would take?

   _____

   _____

   _____

LESSON
2·1 **Estimation Challenge** *continued*

3. **a.** Suppose that you did not stop to rest, eat, sleep, or for any other reason. About how long would it take you to get from school to your destination?

About _____ hours

b. What did you do to estimate how long it would take you?

_____

_____

_____

4. Suppose you start from school at 7:00 A.M. on Monday. You take time out to rest, eat, sleep, and for other reasons.

a. List reasons that you might stop along the way. For each reason, write about how long you would stop.

Reason for Stopping	Length of Stop

b. At about what time, and on what day of the week, would you expect to reach your destination?

Time: About _____ Day: _____

5. Who did you work with on this Estimation Challenge? _____

LESSON 2·1 Math Boxes

1. Solve.

    a.   8.83
        + 3.29

    b.   154.3
        − 28.5

SRB 34–36

2. Make the following changes to the numeral 6,205.12.
Change the digit
in the ones place to 7,
in the hundreds place to 5,
in the tenths place to 6,
in the tens place to 8,
in the thousands place to 4.
Write the new numeral.

—,———.——

SRB 4 28–30

3. Round each number to the nearest thousand.

a. 43,802 _____

b. 904,873 _____

c. 1,380,021 _____

d. 5,067 _____

e. 20,503 _____

SRB 227

4. Write an equivalent decimal.

a. $\frac{1}{2}$ = _____

b. $\frac{3}{4}$ = _____

c. 35% = _____

d. 73% = _____

SRB 50 83

5. Circle the acute angles.

a.   b.

c.   d.

SRB 139

6. Sam drew a trapezoid and a square and covered them as shown. Write the name below each figure. Then finish each drawing.

SRB 146

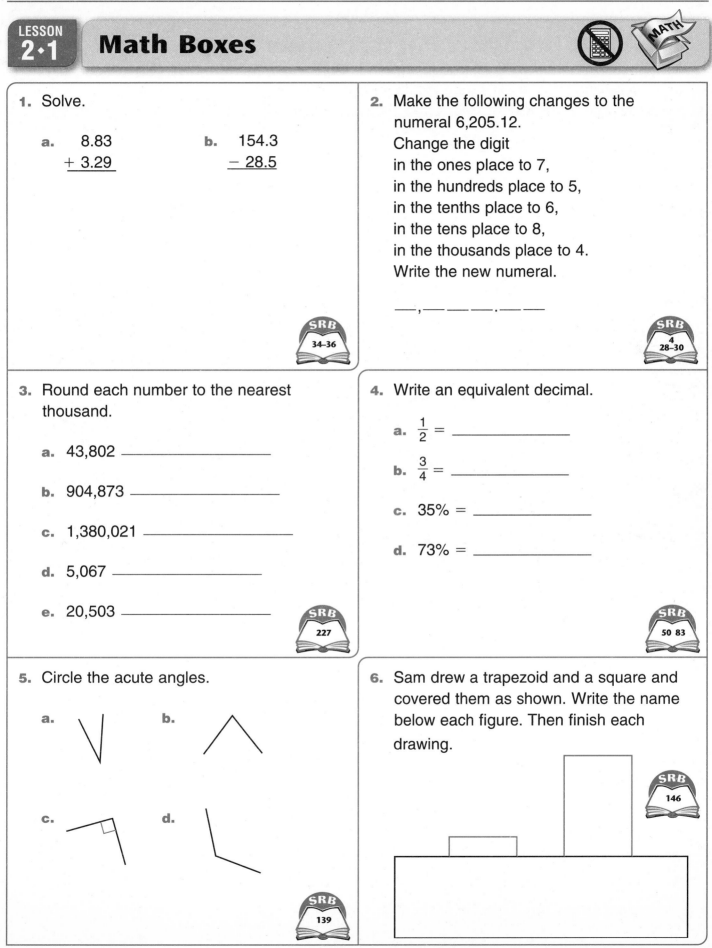

LESSON 2·2   Adding with Partial Sums

Write the following numbers in expanded notation.

1. 432: _____

2. 56.23 _____

Write an estimate for each problem. Then use the partial-sums method to find the exact answer.

Example:

Estimate: _____ 400 _____

```
 325.022
+134.527
+400.000
  50.000
   9.000
   0.500
   0.040
   0.009
 459.549
```

3. Estimate: _____

```
   214
+ 475.2
```

4. Estimate: _____

```
  10.31
  32.04
+ 59.61
```

5. Estimate: _____

```
  28.765
+ 31.036
```

6. Estimate: _____

```
  47.84
+ 21.023
```

LESSON 2·2 Methods for Addition

Solve Problems 1–5 using the partial-sums method. Solve the rest of the problems using any method you choose. Show your work in the space on the right. Compare answers with your partner. If there are differences, work together to find the correct solution.

1. $714 + 465 =$ _____

2. $253 + 187 =$ _____

3. _____ $= 5{,}312 + 3{,}687$

4. $3{,}416 + 2{,}795 =$ _____

5. $475 + 139 + 115 =$ _____

6. _____ $= 217 + 192 + 309 + 536$

7. $38.47 + 9.58 =$ _____

8. _____ $= 32.06 + 65.1$

9. $43.46 + 7.1 + 2.65 =$ _____

10. Alana is in charge of the class pets. She spent
    $ 43.65 on hamster food,
    $ 37.89 on rabbit food,
    $ 2.01 on turtle food, and
    $ 7.51 on snake food.

    How much did she spend on pet food?

    Estimate: _____

    Solution: _____

Math Boxes

1. Round to the nearest tenth.

   a. 45.52 = _____

   b. 60.18 = _____

   c. 123.45 = _____

   d. 38.27 = _____

   e. 56.199 = _____

   SRB
   46

2. Multiply.

   a. 7 * 10 = _____

   b. 7 * 60 = _____

   c. 70 * 60 = _____

   d. 8 * 10 = _____

   e. 8 * 70 = _____

   f. 80 * 70 = _____

   SRB
   18

3. The temperature at midnight was 25°F. The wind chill temperature was 14°F. How much warmer was the actual temperature than the wind chill temperature?

   _____

   SRB
   15–17
   203

4. Complete.

   a. 354 = 300 + 50 + _____

   b. 867 = 800 + _____ + 7

   c. 975 = _____ + 70 + 5

   d. 1,256 = 1,000 + _____ + ____ + 6

   e. 6,704 = 6,000 + _____ + 4

   SRB
   13

5. Complete.

   a. A person 74 in. tall is ____ ft ____ in.

   b. A person who runs a mile runs

      _____ ft.

   c. A person who runs 1,760 yd runs

      _____ ft.

   d. A person who grew $\frac{1}{2}$ ft over

      the summer grew ____ in.

   SRB
   184

6. Match.

   a. Straight angle   <90°

   b. Obtuse angle   >90°

   c. Right angle   90°

   d. Acute angle   180°

   SRB
   139

LESSON 2·3 Methods for Subtraction

Solve Problems 1 and 2 using the trade-first method. Solve Problems 3 and 4 using the partial-differences method. Solve the rest of the problems using any method you choose. Show your work in the space below. Compare your answers with your partner's answers. If there are differences, work together to find the correct solution.

1. $67 - 39 =$ _____

2. _____ $= \$34.68 - \$15.75$

3. $895 - 327 =$ _____

4. $7,053 - 2,690 =$ _____

5. $146.9 - 92.5 =$ _____

6. _____ $= 138.2 - 79.6$

7. _____ $= 5,829 - 673$

8. $9.6 - 4.87 =$ _____

LESSON 2·3

Math Boxes

1. Solve.

   a.   88.5
       + 32.9

   b. $4.48
       − $3.82

   SRB 35 36

2. Make the following changes to the numeral 76.432.
   Change the digit
   in the ones place to 1,
   in the thousandths place to 4,
   in the tenths place to 2,
   in the tens place to 8.
   Write the new numeral.

   _____

   SRB 4 28–30

3. Round each number to the nearest ten-thousand.

   a. 1,308,799 _____

   b. 621,499 _____

   c. 8,003,291 _____

   d. 158,005 _____

   e. 2,226,095 _____

   SRB 249

4. Write an equivalent decimal.

   60% = _____

   $\frac{1}{5}$ = _____

   $\frac{4}{10}$ = _____

   93% = _____

   SRB 50 83

5. Circle the obtuse angles.

   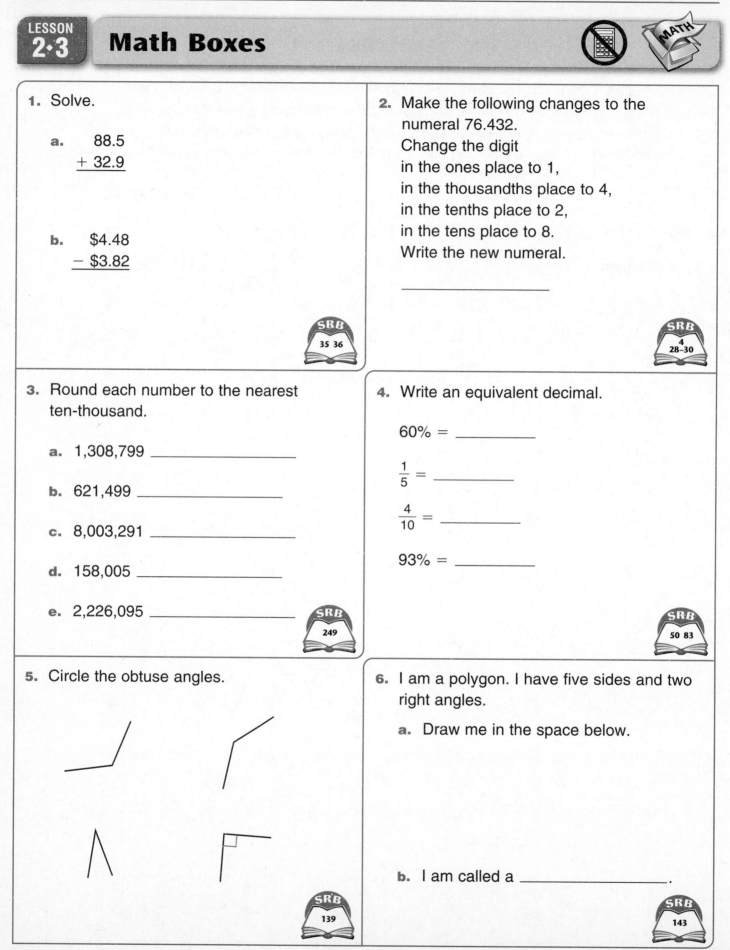

   SRB 139

6. I am a polygon. I have five sides and two right angles.

   a. Draw me in the space below.

   b. I am called a _____.

   SRB 143

LESSON 2·4 Addition and Subtraction Stories

For each problem on pages 37 and 38, fill in the blanks and solve the problem.

Example: Maria had 2 decks of cards. One of the decks had only 36 cards instead of 52. The rest were missing from the deck. How many cards were missing?

◆ List the numbers needed to solve the problem. _____*36 and 52*_____

◆ Describe what you want to find. _____*The number of missing cards*_____

◆ Write an open sentence: _____*36 + c = 52*_____

◆ Find the missing number in the open sentence. Solution: _____*c = 16*_____

◆ Write the answer to the number story. Answer: _____*16 cards*_____
(unit)

1. Anthony got a new bike. He rode 18 miles the first week, 27 miles the second week, and 34 miles the third week. How many miles did he ride altogether?

   a. List the numbers needed to solve the problem. _____

   b. Describe what you want to find. _____

   _____

   c. Open sentence: _____

   d. Solution: _____   e. Answer: _____
   (unit)

2. Regina has $23.08. Her sister has $16.47. Her brother has only $5.00. How much more money does Regina have than her sister?

   a. List the numbers needed to solve the problem. _____

   b. Describe what you want to find. _____

   _____

   c. Open sentence: _____

   d. Solution: _____   e. Answer: _____

3. Lucas was making breakfast for 12 friends. He started with 19 eggs. He bought 1 dozen more eggs. How many eggs did he have to cook for breakfast?

   a. List the numbers needed to solve the problem. _____

   b. Describe what you want to find. _____

   _____

   c. Open sentence: _____

   d. Solution: _____   e. Answer: _____
   (unit)

LESSON 2·4   Addition and Subtraction Stories *continued*

4. Nicholas earned $48 mowing lawns one weekend. With the money he earned, he bought 2 CDs that cost a total of $23. How much money did he have left?

   a. List the numbers needed to solve the problem. _____

   b. Describe what you want to find. _____

   c. Open sentence: _____

   d. Solution: _____   e. Answer: _____

   (unit)

Circle the open sentence that best matches each story and then solve the problem.

5. Patrick paints color-by-number pictures. He spent 24 hours painting in June and 37 hours painting in July. One picture had 18 different colors. How many hours did he paint in the two months?

   $18 + h = 37$ $24 + h = 37$

   $37 + 24 = h$ $37 - h = 18$

   Answer: _____
   (unit)

6. Sue walked 2 miles to Jan's house. Then both girls walked 2 miles to Tad's house. Sue took 28 minutes to get to Jan's house. The girls took 45 minutes to get to Tad's house. How much longer did it take to get to Tad's than to Jan's?

   $2 * 28 = m$ $2 + 28 + m = 45$

   $m - 28 = 45$ $45 - 28 = m$

   Answer: _____
   (unit)

Circle the number sentence that matches the problem.

7. Ralph and Adeline saved their money for 6 weeks. Ralph saved $27.23. Adeline saved $34.98. How much more did Adeline save than Ralph?

   a. $27.23 + $34.98 = s

   b. 6 * s = $34.98

   c. $27.23 + s = $34.98

   d. 6 + s = $34.98 − $27.23

   Answer: _____
   (unit)

8. Tanya's guinea pig is 5 years old and 10.25 inches long. Monique's guinea pig is 2 years old and 11.5 inches. Jamal's guinea pig is 4 years old and 9.3 inches. What is the difference in length between the shortest and the longest guinea pig?

   a. $11.5 - 9.3 = d$ b. $11.5 - 10.25 = d$

   c. $d = 2 + 5 + 9.3$ d. $d = 11.5 + 9.3$

   Answer: _____
   (unit)

LESSON 2·4   Math Boxes

1. Round to the nearest hundredth.

   a. 67.467 = _____

   b. 9.017 = _____

   c. 43.284 = _____

   d. 16.107 = _____

   e. 5.658 = _____

   SRB 46

2. Multiply.

   a. 60 * 4 = _____

   b. 60 * 40 = _____

   c. 60 * 400 = _____

   d. 60 [10s] = _____

   e. 600 [10s] = _____

   f. 600 [100s] = _____

   SRB 18

3. At the start of a science experiment, the temperature in a box was 27°C. The temperature increased by 32 degrees. Then it decreased by 43 degrees. What was the final temperature in the box?

_____

   SRB 15–17 203

4. Complete.

   a. 657.46 = 600 + _____ + 7 + 0.4 + 0.06

   b. 25.72 = 20 + 5 + _____ + 0.02

   c. 94.257 = 90 + _____ + 0.2 +

   _____ + 0.007

   d. 365.27 = _____ + _____

   + _____ + 0.2 + _____

   SRB 13

5. Convert each measurement.

   a. 100 yd = _____ ft

   b. 4 mi = _____ ft

   c. _____ in. = 2 yd

   d. _____ ft = 30 in.

   e. $2\frac{1}{2}$ yd = _____ in.

   SRB 184

6. Identify each angle as *right, acute,* or *obtuse.*

   A: _____

   B: _____

   C: _____

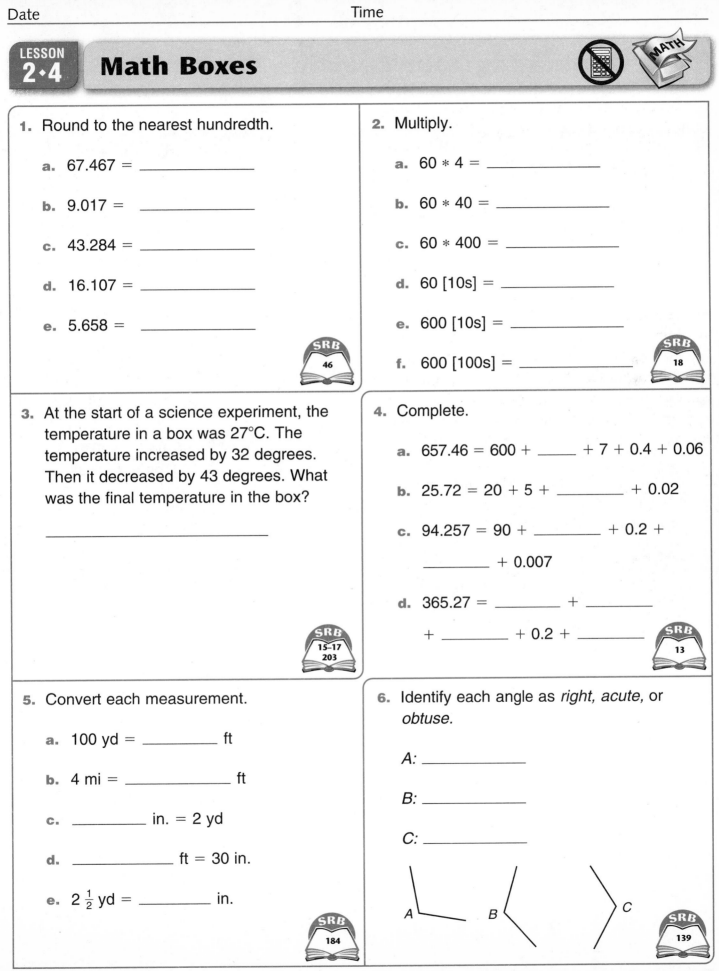

   SRB 139

39

LESSON 2·5 Estimating Your Reaction Time

Tear out Activity Sheet 2 from the back of your journal. Cut out the Grab-It Gauge.

It takes two people to perform this experiment. The "Tester" holds the Grab-It Gauge at the top. The "Contestant" gets ready to catch the gauge by placing his or her thumb and index finger at the bottom of the gauge, *without quite touching it.* (*See diagram.*)

When the Contestant is ready, the Tester lets go of the gauge. The Contestant tries to grab it with his or her thumb and index finger as quickly as possible.

The number grasped by the Contestant shows that person's reaction time, to the nearest hundredth of a second. The Contestant then records that reaction time in the data table shown below.

Partners take turns being Tester and Contestant. Each person should perform the experiment 10 times with each hand.

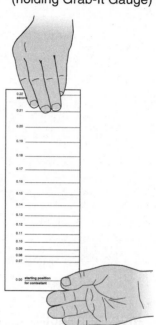

Tester
(holding Grab-It Gauge)

Contestant
(not quite touching
Grab-It Gauge)

Reaction Time (in seconds)			
Left Hand		Right Hand	
1.	6.	1.	6.
2.	7.	2.	7.
3.	8.	3.	8.
4.	9.	4.	9.
5.	10.	5.	10.

LESSON 2·5 **Estimating Your Reaction Time** *continued*

Use the results of your Grab-It experiment to answer the following questions.

1. What was the **maximum** reaction time for your

   left hand? _____ right hand? _____

2. What was the **minimum** reaction time for your

   left hand? _____ right hand? _____

3. What was the **range** of reaction times for your

   left hand? _____ right hand? _____

4. What reaction time was the **mode** for your

   left hand? _____ right hand? _____

5. What was the **median** reaction time for your

   left hand? _____ right hand? _____

6. What was the **mean** reaction time for your

   left hand? _____ right hand? _____

7. If you could use just one number to estimate your reaction time, which number would you choose as the best estimate? Circle one.

   minimum   maximum   mode   median   mean

   Explain. _____

   _____

   _____

8. Which of your hands reacted more quickly in the Grab-It experiment?

   _____

LESSON 2·5 Math Boxes

1. Write each number in exponential notation.

a. $4 * 4 * 4 =$ _____

b. $5 * 5 * 5 * 5 =$ _____

c. $9 * 9 * 9 * 9 =$ _____

d. $7 * 7 =$ _____

e. $2 * 2 * 2 * 2 * 2 =$ _____

SRB 4

2. Estimate. $589 * 714$

a. Write a number sentence:

_____

b. How I estimated. _____

_____

_____

SRB 18 219 247–249

3. Add.

a.
$$632$$
$$+ \ 859$$

b.
$$3,341$$
$$+ \ \ 799$$

SRB 13 14

4. Subtract.

a.
$$1,924$$
$$- \ \ 385$$

b.
$$1,493$$
$$- \ \ 208$$

SRB 15–17

5. Solve.

a.
$$128.07$$
$$- \ 85.25$$

b.
$$306.85$$
$$+ 216.96$$

SRB 34–36

6. Tim flipped a coin 10 times. It landed heads up 7 times and tails up 3 times. Tim said, "I'll flip it 4 more times to get the same number of heads and tails." Is he right? Explain why or why not.

_____

_____

_____

SRB 129

LESSON 2·6   **Describing Chances**

1. Circle the number that best describes the chance of landing in the blue area.

Spinner	Chance of Landing on Blue

a.   0.25   50% $\frac{2}{3}$ 0.75   90%

b.   0.25   50% $\frac{2}{3}$ 0.75   90%

c.   0.25   50% $\frac{2}{3}$ 0.75   90%

d.   0.25   50% $\frac{2}{3}$ 0.75   90%

2. Use the words and phrases from the Word Bank. Write how you would describe the chance of the event happening or not happening.

Word Bank			
certain	extremely likely	very likely	50–50 chance
impossible	extremely unlikely	very unlikely	unlikely likely

Example: Most people will fly in an airplane at least once during their lifetime.

extremely likely

Event	Chance
a. The sun will rise tomorrow.	_____
b. An adult is able to swim.	_____
c. A newborn baby will be a girl.	_____
d. A long-distance call will result in a busy signal.	_____
e. There will be an earthquake in California during the next year.	_____
f. A house in the United States, will catch on fire during the next year.	_____

LESSON 2·6 | # A Thumbtack Experiment

Make a guess: If you drop a thumbtack, is it more likely
to land with the point up or with the point down? _____

The experiment described below will enable you to make an estimate of the
chance that a thumbtack will land point down.

1. Work with a partner. You should have 10 thumbtacks and 1 small cup. Do the
   experiment at your desk or a table so you are working over a smooth, hard surface.

   Place the 10 thumbtacks inside the cup. Shake the cup a few times, and then
   carefully drop the tacks onto the desk surface. Record the number of thumbtacks
   that land point up and the number that land point down.

   Toss the 10 thumbtacks 9 more times and record the results each time.

Toss	Number Landing Point Up	Number Landing Point Down
1		
2		
3		
4		
5		
6		
7		
8		
9		
10		
	Total Up =	Total Down =

2. In making your 10 tosses, you dropped a total of 100 thumbtacks.

   What fraction of the thumbtacks landed point down? _____

3. Write this fraction on a small stick-on note. Also write it as a decimal and as a
   percent.

4. For the whole class, the chance that a tack will land point down is _____.

LESSON 2·6   Magnitude Estimates for Addition and Subtraction

Make a magnitude estimate. Then write a number model to show how you estimated.

1. The air distance from Seattle to Minneapolis is 1,390 miles. The air distance from Minneapolis to New York City is 1,020 miles. What is the air distance from Seattle to New York City, making one stop in Minneapolis?

10s	100s	1,000s	10,000s	100,000s	1,000,000s

Number Model: _____

2. Alaska, the largest state, has an area of 615,230 square miles. Rhode Island, the smallest state, has an area of 1,281 square miles. What is the difference in area?

10s	100s	1,000s	10,000s	100,000s	1,000,000s

Number Model: _____

3. In 2000, George W. Bush received 49,820,518 votes. In 1824, John Quincy Adams received 108,740 votes. How many more votes were cast for Bush than Adams?

10s	100s	1,000s	10,000s	100,000s	1,000,000s	10,000,000s

Number Model: _____

4. A weather station reported rainfall totals for a nearby town over a three-month period. The recorded rainfall was 1.739 inches, 2.067 inches, and 3.987 inches. What was the total rainfall?

0.1s	1s	10s	100s	1,000s

Number Model: _____

5. Choose three of the above problems, and use the space below to calculate a solution.

LESSON 2·6

Math Boxes

1. Give the value of the **boldface** digit in each numeral.

   a. **2**87,051 _____

   b. 7,04**2**,690 _____

   c. **2**8,609,381 _____

   d. 506,344,**5**26 _____

   e. **4**7,381,296 _____

   SRB 4

2. Solve.

   a. $3 + n = 17$ b. $35 - r = 10$

      $n =$ _____ $r =$ _____

   c. $67 + t = 113$ d. $5.9 - b = 2$

      $t =$ _____ $b =$ _____

   e. $3.25 + n = 12.75$

      $n =$ _____

   SRB 219

3. Write the prime factorization of 32.

   _____

   _____

   SRB 12

4. Multiply.

   $30 * 900 =$ _____

   $400 *$ _____ $= 40,000$

   $800 * 6,000 =$ _____

   $2,000 * 500 =$ _____

   _____ $= 600 * 700$

   SRB 18

5. Measure angle *TAG* to the nearest degree.

   Angle *TAG*: _____

   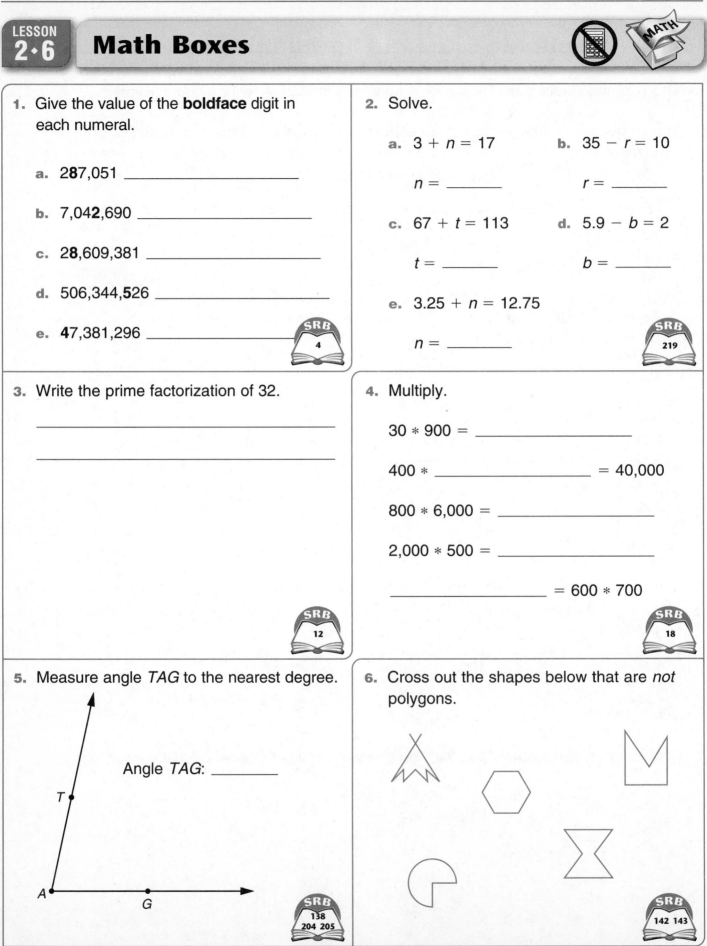

   SRB 138 204 205

6. Cross out the shapes below that are *not* polygons.

   SRB 142 143

46

LESSON 2·7 — Magnitude Estimates for Products

A **magnitude estimate** is a very rough estimate of the answer to a problem. A magnitude estimate will tell you whether the exact answer is in the tenths, ones, tens, hundreds, thousands, and so on.

For each problem, make a magnitude estimate. Ask yourself, "Is the answer in the tenths, ones, tens, hundreds, thousands, or ten-thousands?" Circle the appropriate box. Then write a number sentence to show how you estimated. *Do not solve the problems.*

Example: 14 * 17

10s	⟨100s⟩	1,000s	10,000s

$$10 * 20 = 200$$

How I estimated

---

1. 56 * 37

10s	100s	1,000s	10,000s

How I estimated

---

2. 7 * 326

10s	100s	1,000s	10,000s

How I estimated

---

3. 95 * 48

10s	100s	1,000s	10,000s

How I estimated

---

4. 5 * 4,127

10s	100s	1,000s	10,000s

How I estimated

---

5. 46 * 414

10s	100s	1,000s	10,000s

How I estimated

---

6. 4.5 * 0.6

0.1s	1s	10s	100s

How I estimated

---

7. 7.6 * 9.1

0.1s	1s	10s	100s

How I estimated

---

8. 160 * 2.9

0.1s	1s	10s	100s

How I estimated

---

9. 0.8 * 0.8

0.1s	1s	10s	100s

How I estimated

LESSON 2·7   Spinner Experiments

You can make a spinner by dividing a
circle into different-color parts and
holding a large paper clip in place
with the point of a pencil.

1. Divide the spinner at the right into 3 parts.
   Color the parts red, blue, and green so
   the paper clip has

   ◆ a $\frac{1}{3}$ chance of landing on red;

   ◆ a $\frac{1}{2}$ chance of landing on blue; and

   ◆ a $\frac{1}{6}$ chance of landing on green.

Word Bank			
certain	extremely likely	very likely	50–50 chance
impossible	extremely unlikely	very unlikely	unlikely

Use the words and phrases from the Word Bank.
Describe the chance that the spinner
would land on...

Suppose you spin the paper clip 90 times.
About how many times would you expect
it to land on...

2. red. _____

5. red? _____

3. blue. _____

6. blue? _____

4. green. _____

7. green? _____

8. Spin a paper clip on your spinner 90 times. Tally the results in the table.

Color	Tallies
red	
green	
blue	

9. Did your prediction match your result? Explain on a different piece of paper why you
   think it was the same or different.

LESSON 2·7 — Math Boxes

1. Write the repeated-factor notations.

a. $3^4 = 3 * 3 * 3 * 3$

b. $5^3 =$ _____

c. $7^4 =$ _____

d. $2^5 =$ _____

e. $10^3 =$ _____

SRB
6

2. Estimate. $247 * 974$

a. Write your estimate as a number sentence:

_____

b. How I estimated.

_____

_____

_____

SRB
18 219
247–249

3. Add.

a. $\begin{array}{r} 3{,}672 \\ +\ 1{,}319 \\ \hline \end{array}$

b. $\begin{array}{r} 1{,}654 \\ +\ 2{,}020 \\ \hline \end{array}$

SRB
13 14

4. Subtract.

a. $\begin{array}{r} 322 \\ -\ 199 \\ \hline \end{array}$

b. $\begin{array}{r} 602 \\ -\ 483 \\ \hline \end{array}$

SRB
15–17

5. Solve.

a. $\begin{array}{r} 18.95 \\ -\ 6.07 \\ \hline \end{array}$

b. $\begin{array}{r} 215.29 \\ +\ 38.75 \\ \hline \end{array}$

SRB
34–36

6. When rolling a pair of dice, is there a better chance of rolling a 7 or a 9? Explain.

_____

_____

_____

_____

_____

_____

SRB
129

LESSON 2·8

Multiplication of Whole Numbers

For each problem, make a magnitude estimate. Circle the appropriate box.
Do not solve the problems.

1. 6 * 543

10s	100s	1,000s	10,000s

_____
How I estimated

2. 3 * 284

10s	100s	1,000s	10,000s

_____
How I estimated

3. 46 * 97

10s	100s	1,000s	10,000s

_____
How I estimated

4. 4 * 204

10s	100s	1,000s	10,000s

_____
How I estimated

5. 25 * 37

10s	100s	1,000s	10,000s

_____
How I estimated

6. 56 * 409

10s	100s	1,000s	10,000s

_____
How I estimated

7. Solve each problem above for which your estimate is at least 1,000. Use the partial-products method for at least one problem. Show your work on the grid.

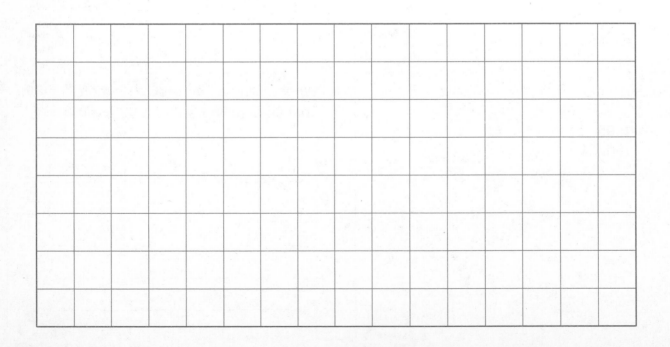

LESSON 2·8 | **Multiplication of Decimals**

For each problem, make a magnitude estimate. Circle the appropriate box.
Do not solve the problems.

1. 2.4 * 63

0.1s	1s	10s	100s

_____
How I estimated

2. 7.2 * 0.6

0.1s	1s	10s	100s

_____
How I estimated

3. 13.4 * 0.3

0.1s	1s	10s	100s

_____
How I estimated

4. 3.58 * 2.1

0.1s	1s	10s	100s

_____
How I estimated

5. 7.84 * 6.05

0.1s	1s	10s	100s

_____
How I estimated

6. 2.8 * 93.6

0.1s	1s	10s	100s

_____
How I estimated

7. Solve each problem above for which your estimate is at least 10. Show your work on the grid below.

Solving Number Stories

For each problem, fill in the blanks and solve the problem.

1. Linell and Ben pooled their money to buy a video game. Linell had $12.40 and Ben had $15.88. How much money did they have in all?

   a. List the numbers needed to solve the problem. _____

   b. Describe what you want to find. _____

   _____

   c. Open sentence: _____

   d. Solution: _____   e. Answer: _____

2. If the video game cost $22.65, how much money did they have left?

   a. List the numbers needed to solve the problem. _____

   b. Describe what you want to find. _____

   _____

   c. Open sentence: _____

   d. Solution: _____   e. Answer: _____

3. Linell and Ben borrowed money so they could also buy a CD for $13.79. How much did they have to borrow so they would have enough money to buy the CD?

   a. List the numbers needed to solve the problem. _____

   b. Describe what you want to find. _____

   _____

   c. Open sentence: _____

   d. Solution: _____   e. Answer: _____

4. How much more did the video game cost than the CD?

   a. List the numbers needed to solve the problem. _____

   b. Describe what you want to find out. _____

   _____

   c. Open sentence: _____

   d. Solution: _____   e. Answer: _____

LESSON 2·8 | **Math Boxes**

1. Use the map on page 386 of your *Student Reference Book* to answer the following questions.

   Choose the best answer.

   a. About how many miles is it from Juneau, Alaska, to the Arctic Circle?

   ⬭ 150 mi   ⬭ 200 mi

   ⬭ 187.5 mi   ⬭ 525 mi

   b. About how far is it from the center of the California-Oregon border to the center of the California-Mexico border?

   ⬭ 1,000 mi   ⬭ 750 mi

   ⬭ 900 mi   ⬭ 350 mi

   SRB 211

2. a. Make up a data set of at least 12 numbers that have the following landmarks.

   maximum: 18   mode: 7

   range: 13   median: 12

   _____

   _____

   b. Make a bar graph of the data.

   SRB 119 122

3. Find the missing numbers and landmarks for the set of numbers below:

   18, 20, 20, 24, 27, 27, _____, 30, 33, 34,

   36, 36, _____

   a. range: 22

   b. mode: 27

   c. minimum: _____

   d. maximum: _____

   SRB 119

4. Acute angles measure greater than 0 degrees and less than 90 degrees. Circle all the acute angles below.

   SRB 139

LESSON 2·9 Lattice Practice

Study the problems and solutions in Column A. Then use lattice multiplication
to solve the problems in Column B.

Column A

Example 1: $5 * 486 = 2{,}430$

10s	100s	1,000s	10,000s

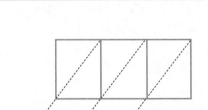

Column B

1. $8 * 274 =$ _____

10s	100s	1,000s	10,000s

Example 2: $87 * 59 = 5{,}133$

10s	100s	1,000s	10,000s

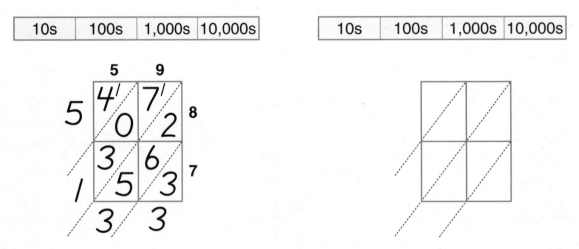

2. $67 * 45 =$ _____

10s	100s	1,000s	10,000s

LESSON 2·9 Multiplication by the Lattice Method

For each problem: ◆ Make a magnitude estimate. Circle the appropriate box.
◆ Solve the problem using the lattice method. Show your work below.

1. 7 * 349 = _____

| 10s | 100s | 1,000s | 10,000s |

2. 48 * 72 = _____

| 10s | 100s | 1,000s | 10,000s |

3. 384 * 256 = _____

| 10s | 100s | 1,000s | 10,000s |

4. 6.15 * 8.3 = _____

| 10s | 100s | 1,000s | 10,000s |

5. 1.7 * 5.6 = _____

| 1s | 10s | 100s | 1,000s |

6. 82 * 4.9 = _____

| 10s | 100s | 1,000s | 10,000s |

LESSON 2·9 Math Boxes

1. Give the value of the **boldface digits** in each numeral.

a. 390.8**1** _____

b. **8**,092,741 _____

c. 4,350.4**7** _____

d. 3**2**,768.9 _____

SRB 4

2. Solve.

a. $n + 45 = 190$

$n =$ _____

b. $360 - n = 270$

$n =$ _____

c. $23.14 + p = 30.59$

$p =$ _____

SRB 219

3. Write the prime factorization of 72.

_____

SRB 12

4. Solve.

a. $8 * 400 =$ _____

b. $36,000 =$ _____ $* 60$

c. $420,000 = 700 *$ _____

d. $9,000 *$ _____ $= 72,000$

e. $5,000 * 8,000 =$ _____

SRB 18 219

5. Measure angle *BOP* to the nearest degree.

∠*BOP:* _____

SRB 138 204 205

6. Cross out the shapes below that are *not* polygons.

SRB 142 143

LESSON 2·10

Millions, Billions, and Trillions

Useful Information	
1 billion is 1,000 times 1 million. 1 million * 1 thousand = 1 billion 1,000,000 * 1,000 = 1,000,000,000	1 trillion is 1,000 times 1 billion. 1 billion * 1 thousand = 1 trillion 1,000,000,000 * 1,000 = 1,000,000,000,000
1 minute = 60 seconds 1 hour = 60 minutes 1 day = 24 hours 1 year = 365 days (366 days in a leap year)	

Make a guess: How long do you think it would take you to tap your desk 1 million times, without any interruptions?

_____

Check your guess by doing the following experiment.

1. Take a sample count.
   Record your count of taps made in 10 seconds. _____

2. Calculate from the sample count.
   At the rate of my sample count, I expect to tap my desk:

a. _____ times in 1 minute.
   (*Hint:* How many 10-second intervals are there in 1 minute?)

b. _____ times in 1 hour.

c. _____ times in 1 day (24 hours).

d. At this rate it would take me about _____ full 24-hour days to tap my desk 1 million times.

3. Suppose that you work 24 hours per day tapping your desk. Estimate how long it would take you to tap 1 billion times and 1 trillion times.

a. It would take me about _____ to tap my desk 1 billion times.
                                              (unit)

b. It would take me about _____ to tap my desk 1 trillion times.
                                              (unit)

LESSON 2·10   **Math Boxes**

1. Find the missing numbers and landmarks for the set of numbers below.

   48, 50, 51, 51, 57, 59, 60, 63, 69, _____, 76, _____

   a. Range: _____

   b. Mode: 76

   c. Minimum: _____

   d. Maximum: 76

   SRB 119

2. a. Make up a set of at least twelve numbers that have the following landmarks.

   Maximum: 8   Range: 6
   Mode: 6   Median: 5

   _____

   _____

   b. Make a bar graph of the data.

   SRB 119 122

3. Use the map on page 354 of your *Student Reference Book* to answer the questions. Choose the best answer.

   a. What is the shortest distance between San Francisco and San Antonio? About:

   ⬭ 500 miles   ⬭ 1,000 miles   ⬭ 1,500 miles   ⬭ 2,000 miles

   b. What is the shortest distance between New York City and Chicago? About:

   ⬭ 700 miles   ⬭ 800 miles   ⬭ 900 miles   ⬭ 1,000 miles

   SRB 211

4. a. Circle the times for which the hands on a clock form an acute angle.

   2:00   6:40   1:30   12:50

   b. Circle the times for which the hands on a clock form an obtuse angle.

   8:00   1:20   5:15   10:30

   SRB 139

Date Time

LESSON 2•11

Math Boxes

1. Measure angle *CAT* to the nearest degree.

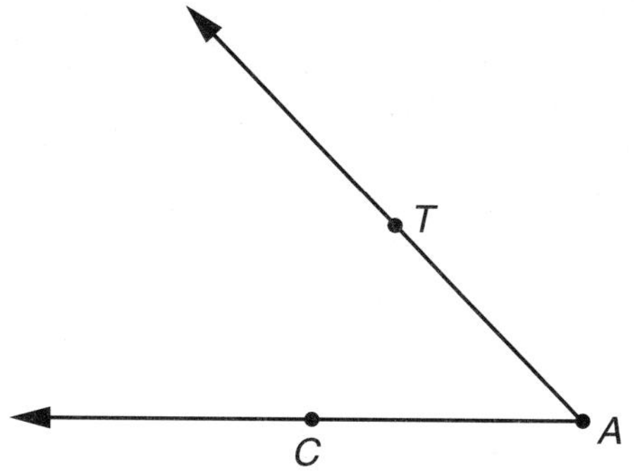

∠*CAT*: ______________

SRB 138 204 205

2. Write the name of an object in the room that has a length of . . .

a. about 30 centimeters.

____

b. about 18 inches.

____

SRB 185

3. For each shape, fill in the ovals that apply.

SRB 142–146 153

a.

- O polygon
- O parallelogram
- O quadrangle
- O rectangle

b.

- O polygon
- O rectangle
- O quadrangle
- O parallelogram

c.

- O polygon
- O triangle
- O circle
- O parallelogram

d.

- O polygon
- O circle
- O quadrangle
- O triangle

4. a. Draw two lines that meet at right angles.

b. What is the measure of each angle?

____

SRB 138 204 205

5. Measure each line segment to the nearest quarter-inch.

a. ________________

________ in.

b. ________

________ in.

SRB 183

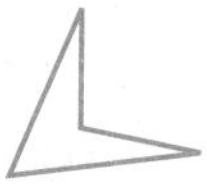

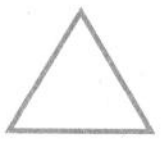

LESSON 3·1 U.S. Census Questions

Use the information on pages 369, 370, 374, and 375 of the *Student Reference Book* to compare the 1790 U.S. Census with the 2000 U.S. Census.

1. a. Which census asked more questions? _____

   b. How many more? _____

2. a. Which census took longer to collect its information? _____

   b. About how much longer did it take?

_____

3. a. In the 1790 census, what percent of households was asked the same questions?

_____

   b. In the 2000 census what percent of households was asked all questions?

_____

4. a. Which state reported the largest total population in the 1790 census? _____

   b. Which state reported the smallest total population in the 1790 census?

   _____

5. What was the reported total population in 1790? _____

_____

_____

6. Do you think it is a good idea that the U.S. Census is done every 10 years? Explain why or why not.

_____

_____

_____

_____

LESSON 3·1   Math Boxes

1. Which triangle is identical to Figure Y?

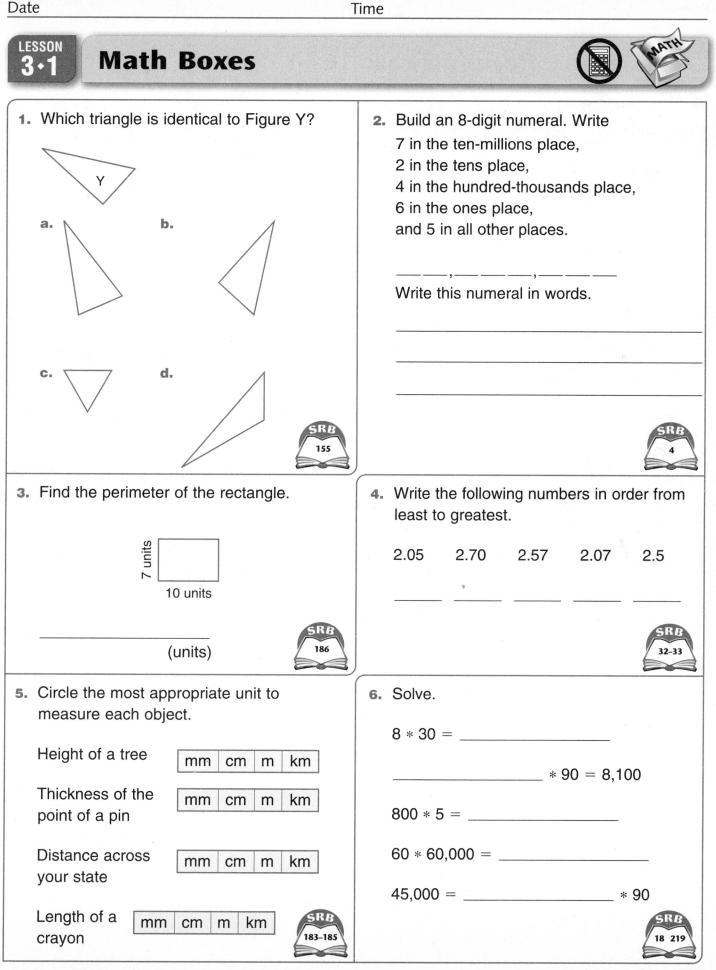

Y

a.

b.

c.

d.

SRB
155

2. Build an 8-digit numeral. Write

7 in the ten-millions place,
2 in the tens place,
4 in the hundred-thousands place,
6 in the ones place,
and 5 in all other places.

___ ___ , ___ ___ ___ , ___ ___ ___

Write this numeral in words.

_____

_____

_____

SRB
4

3. Find the perimeter of the rectangle.

7 units

10 units

_____ (units)

SRB
186

4. Write the following numbers in order from least to greatest.

2.05   2.70   2.57   2.07   2.5

_____ _____ _____ _____ _____

SRB
32–33

5. Circle the most appropriate unit to measure each object.

Height of a tree

| mm | cm | m | km |

Thickness of the point of a pin

| mm | cm | m | km |

Distance across your state

| mm | cm | m | km |

Length of a crayon

| mm | cm | m | km |

SRB
183–185

6. Solve.

$8 * 30 =$ _____

_____ $* 90 = 8{,}100$

$800 * 5 =$ _____

$60 * 60{,}000 =$ _____

$45{,}000 =$ _____ $* 90$

SRB
18 219

LESSON 3·2   State Populations, 1610–1790

Use the population table on *Student Reference Book,* page 371, to answer
the following.

1. What was the population of Pennsylvania in 1780? _____

2. What was the total population of all states in 1760? _____

3. a. Which colony started with the smallest population?

   Name of colony _____

   Year _____

   Population _____

   b. What was the population of this state in the census of 1790? _____

4. Which colony was the first to have a population of more than 100,000?

   Name of colony _____

   Year _____

   Population _____

5. a. In what year was the total population of all states greater than 1 million

   for the first time? _____

   b. In what year was the total population of all states greater than 2 million

   for the first time? _____

6. In 1790, which state had the largest population?

   Name of state _____

   Population _____

LESSON 3·2

State Populations, 1610–1790 *continued*

7. In 1790, which states had smaller populations than Rhode Island?

_____

8. Below, fill in the total U.S. populations for 1780 and 1790. Then find how much the
   population increased during that 10-year period.

   Population in 1790 _____

   Population in 1780 _____

   Increase _____

Try This

9. The table gives the population of Connecticut in 1750 as 100,000. Make a mark in
   front of the statement below that best describes the population of Connecticut in 1750.

   _____ It was exactly 100,000.

   _____ It was most likely between 99,000 and 101,000.

   _____ It was most likely between 95,000 and 105,000.

   Explain your answer.

   _____

   _____

   _____

   _____

   _____

   _____

   _____

   _____

LESSON 3·2 Addition and Subtraction Number Stories

For each problem, fill in the blanks and solve the problem.

1. Jeanne practiced her multiplication facts for 3 weeks. The first week she practiced for 45 minutes, the second week for 37 minutes, and the third week for 32 minutes. How many minutes did she practice in all?

   a. List the numbers needed to solve the problem. _____

   b. Describe what you want to find. _____

   _____

   c. Open sentence: _____

   d. Solution: _____   e. Answer: _____

   (unit)

2. The shortest book Martha read one summer was 57 pages. The longest book was 243 pages. She read a total of 36 books. How many pages longer was the longest book than the shortest book?

   a. List the numbers needed to solve the problem. _____

   b. Describe what you want to find. _____

   _____

   c. Open sentence: _____

   d. Solution: _____   e. Answer: _____

   (unit)

3. Cezar collects marbles. He had 347 marbles. Then he played in two tournaments. He lost 34 marbles in the first tournament. He won 23 marbles in the second tournament. How many marbles did he have after playing in both tournaments?

   a. List the numbers needed to solve the problem. _____

   b. Describe what you want to find. _____

   _____

   c. Open sentence: _____

   d. Solution: _____   e. Answer: _____

   (unit)

LESSON 3·2   Math Boxes

1. Estimate and solve.

   289
  +245

Estimate: _____

Solution: _____

   1,013
  − 867

Estimate: _____

Solution: _____

SRB 13–17

2. Find the landmarks for this set of numbers: 273, 280, 298, 254, 328, 269, 317, 280, 309

Maximum: _____

Minimum: _____

Range: _____

Median: _____

SRB 119

3. Solve.

$5 * m = 45$ $m = $ _____

$8 = 64 ÷ d$ $d = $ _____

$8 = 48 ÷ k$ $k = $ _____

$40 * s = 280$ $s = $ _____

$w * 90 = 54,000$ $w = $ _____

SRB 219

4. Estimate and solve.

a. $42.346 + 37.987$

Estimate: _____

Solution: _____

b. $71.643 − 29.846$

Estimate: _____

Solution: _____

SRB 34–36

5. Write the name of an object in the room that is about 15 centimeters long.

_____

Write the name of an object in the room that is about 3 inches long.

_____

SRB 185

6. Solve.

$76 * 38 = $ _____

$3.7 * 46 = $ _____

$247 * 32 = $ _____

$0.5 * 43.1 = $ _____

$65.2 * 5.7 = $ _____

SRB 19 20 38–40

LESSON 3·3 — Pattern-Block Angles

For each pattern block below, tell the degree measure of the angle and explain how you found the measure. Do not use a protractor.

1.

measure of $\angle A$ = _____

Explain. _____

_____

_____

_____

2.

$m\angle B$ = _____ ($m\angle B$ means measure of angle B.)

Explain. _____

_____

_____

_____

_____

3.

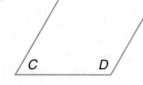

$m\angle C$ = _____ $m\angle D$ = _____

Explain. _____

_____

_____

_____

_____

4.

$m\angle E$ = _____ $m\angle F$ = _____

Explain. _____

_____

_____

_____

_____

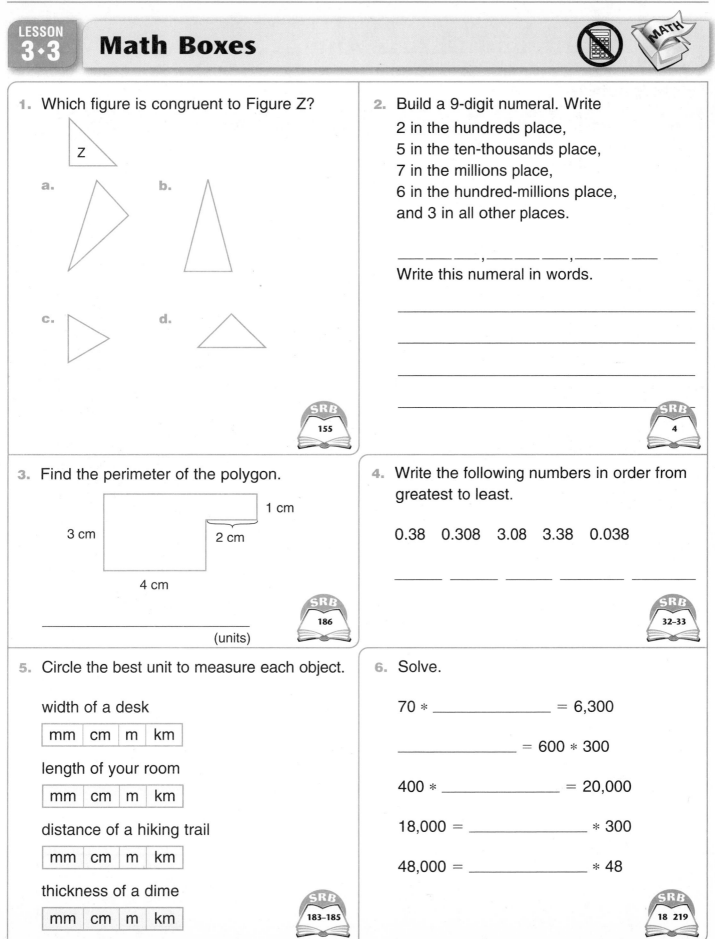

LESSON 3·3

Math Boxes

1. Which figure is congruent to Figure Z?

   Z

   a.

   b.

   c.

   d.

   SRB 155

2. Build a 9-digit numeral. Write

   2 in the hundreds place,
   5 in the ten-thousands place,
   7 in the millions place,
   6 in the hundred-millions place,
   and 3 in all other places.

   ___ ___ ___ , ___ ___ ___ , ___ ___ ___

   Write this numeral in words.

   _____

   _____

   _____

   _____

   SRB 4

3. Find the perimeter of the polygon.

   1 cm
   3 cm
   2 cm
   4 cm

   _____
                                (units)

   SRB 186

4. Write the following numbers in order from greatest to least.

   0.38   0.308   3.08   3.38   0.038

   _____ _____ _____ _____ _____

   SRB 32–33

5. Circle the best unit to measure each object.

   width of a desk

   | mm | cm | m | km |

   length of your room

   | mm | cm | m | km |

   distance of a hiking trail

   | mm | cm | m | km |

   thickness of a dime

   | mm | cm | m | km |

   SRB 183–185

6. Solve.

   $70 * \underline{\hspace{2cm}} = 6{,}300$

   $\underline{\hspace{2cm}} = 600 * 300$

   $400 * \underline{\hspace{2cm}} = 20{,}000$

   $18{,}000 = \underline{\hspace{2cm}} * 300$

   $48{,}000 = \underline{\hspace{2cm}} * 48$

   SRB 18 219

LESSON 3·4 Acute and Obtuse Angles

Math Message

1. **Acute Angles**

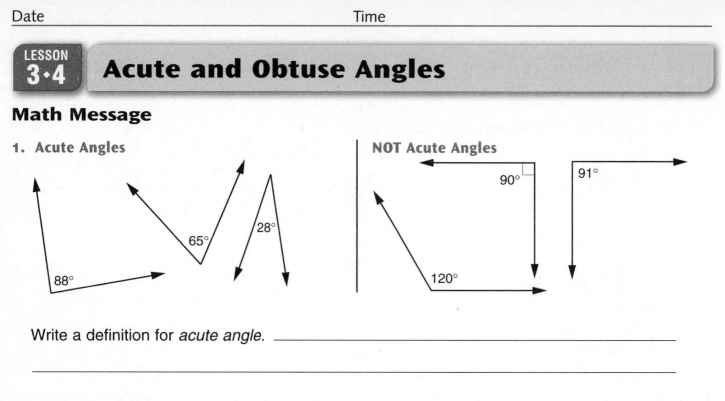

NOT Acute Angles

Write a definition for *acute angle*. _____

_____

2. **Obtuse Angles**

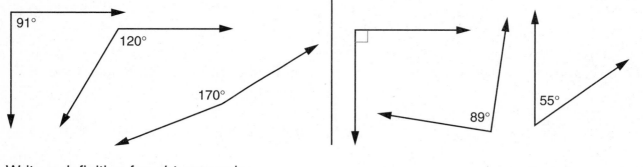

NOT Obtuse Angles

Write a definition for *obtuse angle*. _____

_____

Measuring and Drawing Angles with a Protractor

Sarah used her half-circle protractor to measure the angle at the right. She said it measures about 35°. Theresa measured it with her half-circle protractor. Theresa said it measures about 145°. Devon measured it with his full-circle protractor. Devon said it measures about 325°.

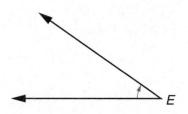

3. **a.** Use both your template protractors to measure the angle. Do you agree with

Sarah, Theresa, or Devon? _____

b. Why? _____

_____

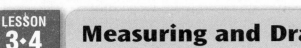

Measuring and Drawing Angles with a Protractor

LESSON 3·4

Use your half-circle protractor. Measure each angle as accurately as you can.

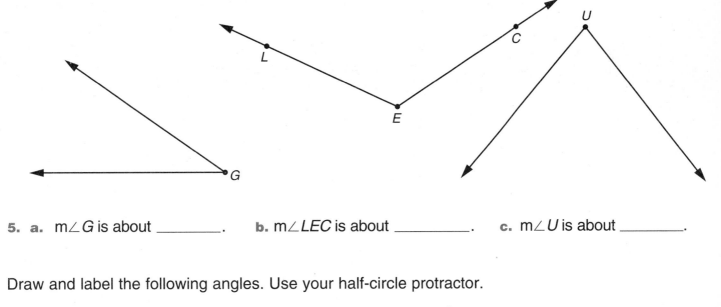

4. **a.** m∠A is about _____.   **b.** m∠EDS is about _____.   **c.** m∠T is about _____.

Use your full-circle protractor to measure each angle.

5. **a.** m∠G is about _____.   **b.** m∠LEC is about _____.   **c.** m∠U is about _____.

Draw and label the following angles. Use your half-circle protractor.

6. **a.** ∠CAT: 62°   **b.** ∠DOG: 135°

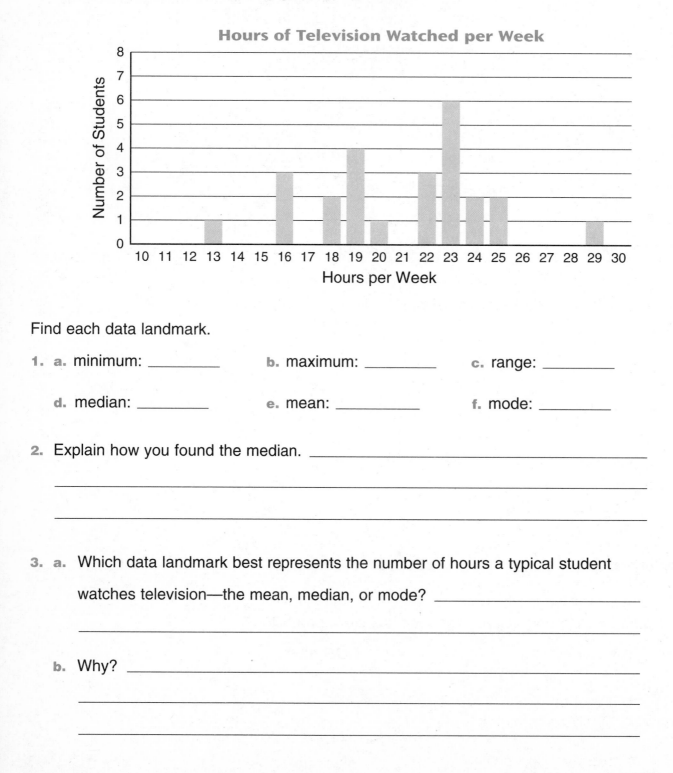

LESSON 3·4

Watching Television

Adeline surveyed the students in her class to find out how much television they watch in a week. She made the following graph of the data.

Hours of Television Watched per Week

Number of Students

Hours per Week

Find each data landmark.

1. a. minimum: _____   b. maximum: _____   c. range: _____

    d. median: _____   e. mean: _____   f. mode: _____

2. Explain how you found the median. _____

_____

_____

3. a. Which data landmark best represents the number of hours a typical student

    watches television—the mean, median, or mode? _____

    _____

    b. Why? _____

    _____

    _____

LESSON 3·4 · Math Boxes

1. Estimate and solve.

    463
  + 2,078

Estimate: _____

Solution: _____

    5,046
  − 2,491

Estimate: _____

Solution: _____

SRB 13–17

2. Find the landmarks for this set of numbers:
99, 87, 85, 32, 57, 82, 85, 99, 85, 65, 78, 87, 85, 57, 85, 99

Maximum: _____

Minimum: _____

Range: _____

Median: _____

SRB 119

3. Solve.

$23 + x = 60$ $x =$ _____

$36 = p * 4$ $p =$ _____

$200 = 50 * m$ $m =$ _____

$55 + t = 70$ $t =$ _____

$28 - b = 13$ $b =$ _____

SRB 219

4. Estimate and solve.

a.   473.894
       + 59.235

Estimate: _____

Solution: _____

b.   78.896
        − 29.321

Estimate: _____

Solution: _____

SRB 34–36

5. Write the name of an object in the room that is about 10 inches long.

_____

Write the name of an object in the room that is about 10 centimeters long.

_____

SRB 185

6. Solve.

$34 * 62 =$ _____

$5.8 * 76 =$ _____

$159 * 7 =$ _____

$0.4 * 231 =$ _____

$76.4 * 8.3 =$ _____

SRB 19 20 38–40

71

LESSON 3·5   Copying Line Segments and Finding Lengths

1. Use your compass and straightedge to copy line segment $\overline{AB}$. Do not measure the line segment with a ruler. Label the endpoints of the new line segment as points M and N. Line segment $\overline{MN}$ should be the same length as line segment $\overline{AB}$.

A •————————————————————————• B

2. Three line segments are shown below:

A •————————————• B   C •———————————————————————• D   E •————————————————• F

Use your compass and straightedge. Construct one line segment that is as long as the three segments joined together end to end. Label the two endpoints of the long line segment X and Y.

Use your compass to find the lengths of different parts of the Geometry Template.

Example: Find the length of the longer side of the rectangle on the Geometry Template.

Step 1 Open the compass to the length of the longer side.

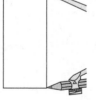

Step 2 Don't change the opening on your compass. Hold the compass against the inch ruler with the anchor at 0. Read the length. The length is about 1 inch.

3. The length of the longer side of the trapezoid is about _____ inch(es).

4. The diameter of the full-circle protractor is about _____ inch(es).

5. The distance between the center of the full-circle protractor and the center of the Percent Circle is about _____ inch(es).

6. Use your compass and a ruler to find two other lengths. Be sure to include units.

Part Measured	Length

LESSON 3·5   Adjacent and Vertical Angles

Angles that are "next to" each other are called **adjacent angles.** Adjacent angles have the same vertex and a common side.

When two lines intersect, four angles are formed. The angles opposite each other are called **vertical angles** or **opposite angles.**

1. **a.** Angles *ABD* and *CBE* are vertical angles.
   Name another pair of vertical angles.

   _____

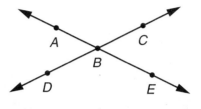

   b. Angles *ABC* and *CBE* are adjacent angles. Name two other pairs of adjacent angles.

   _____

2. The two lines at the right intersect to form
   four angles. One angle has been measured.
   Use your full-circle protractor to measure the
   other three angles. Record your measurements
   on the drawing.

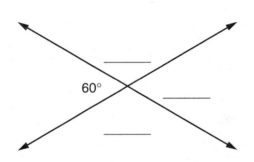

3. On a blank sheet of paper, draw two lines that intersect. Measure the four angles. Record the measures on your drawing.

4. What do you notice about the measures of pairs of vertical angles?

   _____

5. What do you notice about the measures of pairs of adjacent angles?

   _____

Try This

6. For any pair of adjacent angles formed by two intersecting lines, the sum of the measures is

   always 180°. Explain why. _____

   _____

LESSON 3·5 — Math Boxes

1. Measure angle E to the nearest degree.

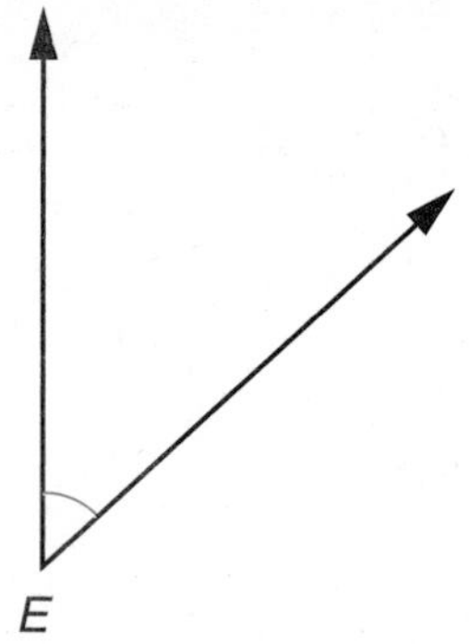

The measure of angle E is about _____.

SRB 204–205

2. Key: ☆ = 1 day absent

Student	Days Absent
Lucca	☆ ☆
Marissa	☆ ☆ ☆
Chandler	☆
Emma	☆ ☆ ☆ ☆

a. Who was absent the most?

_____

b. Who was absent two days?

_____

c. How many days was Chandler absent?

_____

d. Was any student absent more than five days? _____

SRB 117

3. Round 14.762 to the nearest …

tenth. _____

whole number. _____

hundredth. _____

SRB 30 45–46

4. Complete each pattern.

25, _____, 61, _____, 97

87, _____, 43, _____, −1

21, _____, 49, _____, 77

64, _____, _____, _____, 32, 24

61, _____, _____, _____, 81, 86

SRB 230–231

5. List all the factors of 48.

_____

_____

SRB 10

6. Write the prime factorization for 54.

_____

SRB 12

74

LESSON 3·6 Types of Triangles

There are small marks on the sides of some figures below. These marks show sides that are the same length. For example, in the first triangle under Equilateral Triangles, all the sides have two marks. These sides are the same length.

For each type of triangle below, study the examples and nonexamples. Then write your own definitions. Do not use your *Student Reference Book.*

1. Equilateral Triangles

NOT Equilateral Triangles

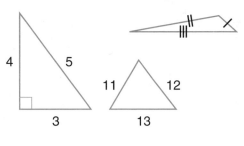

Write a definition of equilateral triangle. _____

_____

_____

2. Isosceles Triangles

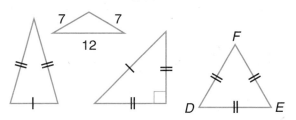

NOT Isosceles Triangles

Write a definition of isosceles triangle. _____

_____

_____

3. Scalene Triangles

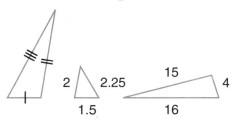

NOT Scalene Triangles

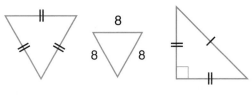

Write a definition of scalene triangle. _____

_____

75

Copying a Triangle

If two triangles are identical—exactly the same size and shape—they are **congruent** to each other. Congruent triangles would match perfectly if you could move one on top of the other.

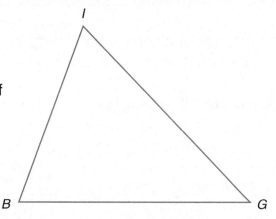

1. **a.** Make a copy of triangle *BIG* on a blank sheet of paper. Use any of your drawing and measuring tools, but DO NOT trace triangle *BIG*. The sides of your copy should be the same length as the sides of triangle *BIG*. The angles also should be the same size as the angles of triangle *BIG*.

   b. When you are satisfied with your work, cut it out and tape it in the space below. Label the vertices *P, A,* and *L*. Triangle *PAL* should be congruent to triangle *BIG*.

LESSON 3·6 Copying More Triangles

1. **a.** Measure the sides of triangle *HOT* in centimeters. Write the lengths next to the sides.

   b. Make a careful copy of triangle *HOT* on a blank sheet of paper. You may use any tools EXCEPT your protractor. DO NOT trace the triangle. When you are satisfied with your work, cut it out and tape it in the space below triangle *HOT*. Label the vertices *R, E,* and *D.*

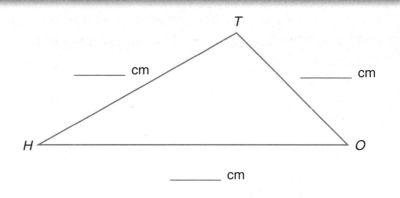

_____ cm

_____ cm

_____ cm

2. Make a copy of triangle *MAX* on a blank sheet of paper.

   Use your compass and straightedge. DO NOT use your ruler or protractor. You may *not* measure the sides. When you are satisfied with your work, cut it out and tape it in the space below triangle *MAX*. Label the vertices *Y, O,* and *U.*

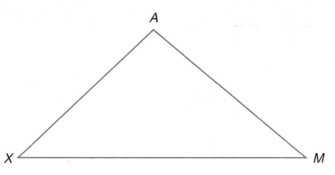

Copying a Partner's Triangle

1. Use a ruler to draw two triangles on a blank sheet of paper. Make your triangles fairly large, but leave enough room to draw a copy of each one. Then exchange drawings with your partner.

2. Copy your partner's triangles using only your compass and straightedge. Don't erase the arcs you make—they show how you made your copies. Measure the sides of the triangles and your copies of the triangles. Write the lengths next to the sides.

3. Cut out one of the triangles your partner drew, and cut out the copy you made. Tape them in the space below.

LESSON 3·6

Math Boxes

1. Write five names for 100,000.

_____

_____

_____

_____

_____

_____

SRB
219

2. Use a straightedge to draw an angle that is greater than 90°.

SRB
139

3. Write < or >.

0.17 _____ 1.71

0.03 _____ 0.12

1.9 _____ 1.89

5.4 _____ 5.04

2.24 _____ 2.2

SRB
9 32
33

4. I am a polygon. I have fewer sides than a quadrangle. Draw me in the space below.

What shape am I? _____

SRB
144

5. What is the measure of angle *T*?

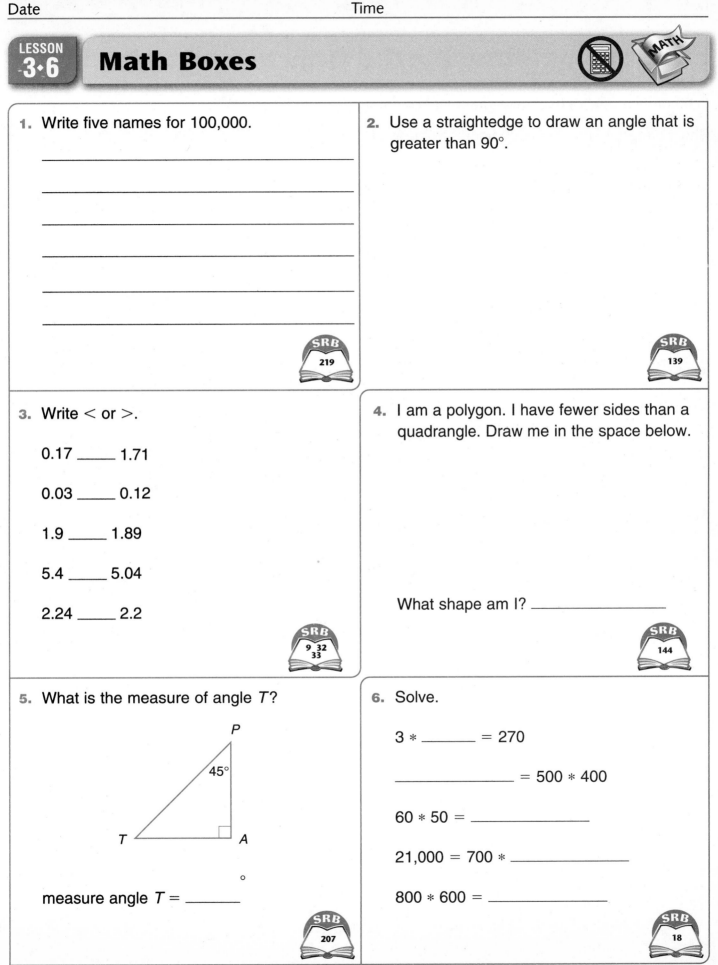

P

45°

T   *A*

°

measure angle *T* = _____

SRB
207

6. Solve.

3 * _____ = 270

_____ = 500 * 400

60 * 50 = _____

21,000 = 700 * _____

800 * 600 = _____

SRB
18

79

Date Time

LESSON 3•7

Completing Partial Drawings of Polygons

Gina drew four shapes: equilateral triangle, square, rhombus, and hexagon.

She covered up most of each figure, as shown below.

Can you tell which figure is which? Write the name below each figure. Then try to draw the rest of the figure.

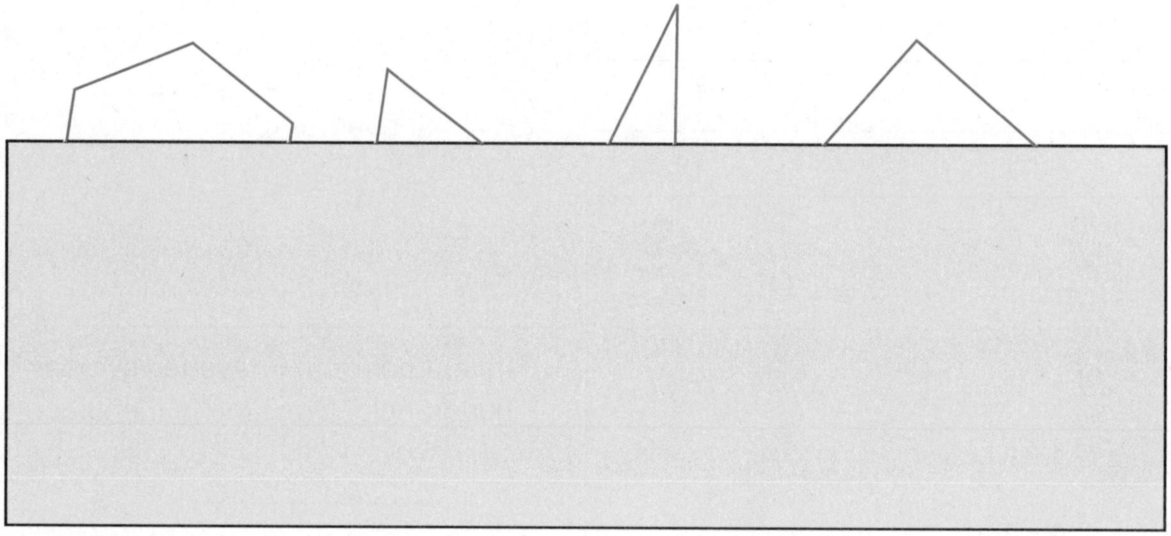

Explain how you solved this problem.

LESSON 3·7 Math Boxes

1. Measure angle *B* to the nearest degree.

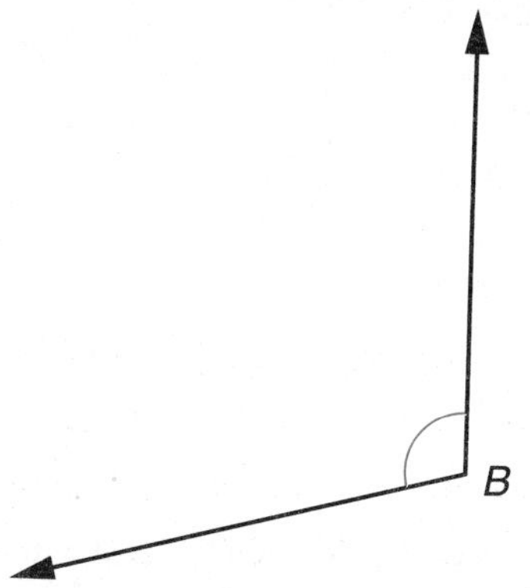

The measure of angle *B* is about _____ °.

SRB 205–206

2. Key: 🎾 = 2 home runs

Player	Home runs
Joe	🎾 🎾 🎾
Yoshi	🎾 🎾
Gregg	🎾 🎾 🎾 🎾
Maria	🎾 🎾 🎾 🎾

a. Who had the most home runs? _____

b. Who had four home runs? _____

c. How many home runs did Maria have?

_____

d. Did any player have fewer than three home runs? _____

SRB 117

3. Round 30.089 to the nearest ...

tenth. _____

whole number. _____

hundredth. _____

SRB 30 45 46

4. Complete each pattern.

17, _____, _____, 62, _____, 92

39, _____, _____, _____, 75, 84

57, _____, _____, 33, _____, 17

15, _____, _____, 33, _____, 45

SRB 230–231

5. List all the factors for 144.

_____

_____

_____

SRB 10

6. Write the prime factorization for 48.

_____

SRB 12

LESSON 3·8 Regular Tessellations

1. A **regular polygon** is a polygon in which all sides are the same length and all angles have the same measure. Circle the regular polygons below.

2. In the table below, write the name of each regular polygon under its picture. Then, using the polygons that you cut out from Activity Sheet 3, decide whether each polygon can be used to create a regular tessellation. Record your answers in the middle column. In the last column, use your Geometry Template to draw examples showing how the polygons tessellate or don't tessellate. Record any gaps or overlaps.

Polygon	Tessellates? (yes or no)	Draw an Example
_____		
_____		
_____		

LESSON 3·8 Regular Tessellations *continued*

Polygon	Tessellates? (yes or no)	Draw an Example
_____		
_____		

3. Which of the polygons can be used to create regular tessellations?

_____

4. Explain how you know that these are the only ones. _____

_____

_____

_____

LESSON 3·8   Math Boxes

1. Circle the name(s) of the shape(s) that could be partially hidden behind the wall.

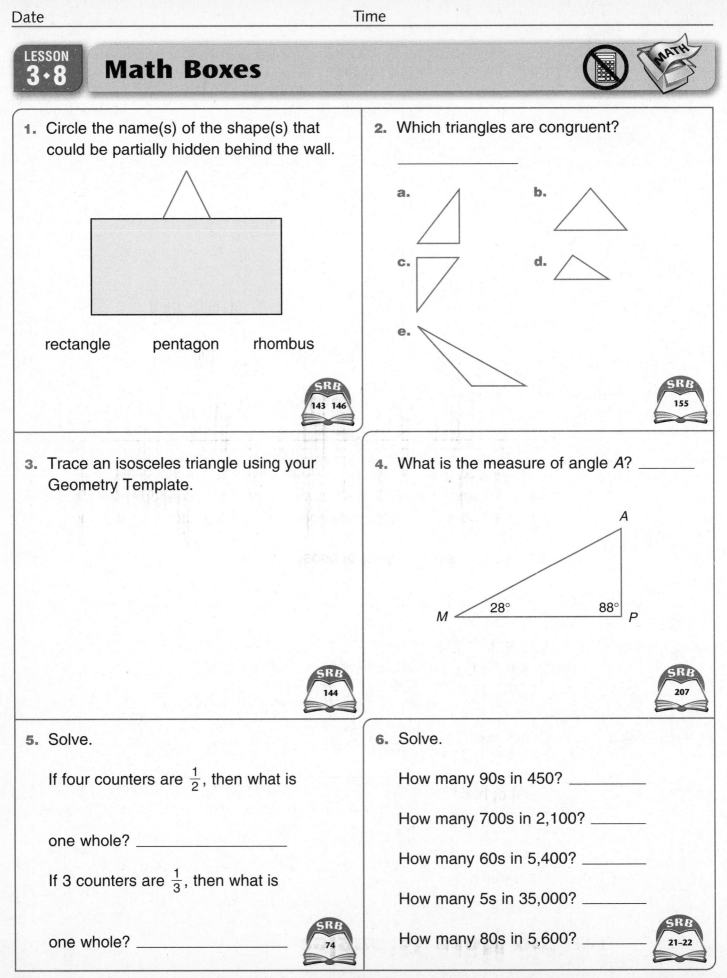

rectangle   pentagon   rhombus

SRB 143 146

2. Which triangles are congruent?

_____

a.

b.

c.

d.

e.

SRB 155

3. Trace an isosceles triangle using your Geometry Template.

SRB 144

4. What is the measure of angle A? _____

SRB 207

5. Solve.

If four counters are $\frac{1}{2}$, then what is

one whole? _____

If 3 counters are $\frac{1}{3}$, then what is

one whole? _____

SRB 74

6. Solve.

How many 90s in 450? _____

How many 700s in 2,100? _____

How many 60s in 5,400? _____

How many 5s in 35,000? _____

How many 80s in 5,600? _____

SRB 21–22

LESSON 3·9   Angles in Quadrangles and Pentagons

1. Circle the kind of polygon your group is working on: quadrangle   pentagon

2. Below, use a straightedge to carefully draw the kind of polygon your group is working on. Your polygon should look different from the ones drawn by others in your group, but it should have the same number of sides.

3. Measure the angles in your polygon. Write each measure for each angle.

_____

4. Find the sum of the angles in your polygon. _____

LESSON 3·9

Angles in Quadrangles and Pentagons *cont.*

5. Record your group's data below.

Group Member's Name	Sketch of Polygon	Sum of Angles

6. Find the median of the angle sums for your group. _____

7. If you have time, draw a hexagon. Measure its angles with a protractor. Find the sum.

Sum of the angles in a hexagon = _____

LESSON 3·9 **Angles in Quadrangles and Pentagons** *cont.*

8. Record the class data below.

Sum of the Angles in a Quadrangle	
Group	Group Median

Sum of the Angles in a Pentagon	
Group	Group Median

9. Find the class median for each polygon. For the triangle, use the median from the Math Message.

Sums of Polygon Angles	
Polygon	Class Median
triangle	
quadrangle	
pentagon	
hexagon	

10. What pattern do you see in the Sums of Polygon Angles table?

_____

_____

_____

LESSON 3·9 **Angles in Heptagons**

1. A heptagon is a polygon with 7 sides.

   Predict the sum of the angles in a heptagon. _____

2. Draw a heptagon below. Measure its angles with a protractor. Write each measure in the angle. Find the sum.

   Sum of the angles in a heptagon = _____

3. **a.** Is your measurement close to your prediction? _____

   b. Why might your prediction and your measurement be different?

   _____

   _____

   _____

   _____

LESSON 3·9 Angles in Any Polygon

1. Draw a line segment from vertex *A* of this octagon to each of the other vertices except *B* and *H*.

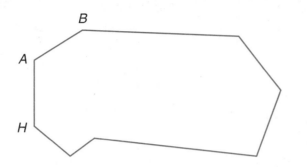

2. How many triangles did you divide the octagon into? _____

3. What is the sum of the angles in this octagon? _____

4. Ignacio said the sum of his octagon's angles is 1,440°. Below is the picture he drew to show how he found his answer. Explain Ignacio's mistake.

_____

_____

_____

_____

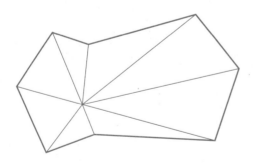

5. A 50-gon is a polygon with 50 sides. How could you find the sum of the angles

in a 50-gon? _____

_____

_____

Sum of the angles in a 50-gon = _____

LESSON 3·9   # Practicing Expanded Notation   SRB 396

Use the place-value chart on page 396 of the *Student Reference Book* to help you write the following numbers in expanded notation.

1. 6,456 = _____

2. 64.56 = _____

3. 98,204 = _____

4. 982.04 = _____

5. a. Build a 4 digit numeral. Write

   3 in the hundredths place,
   4 in the tens place,
   6 in the ones place, and
   9 in the tenths place.

   ____ ____ ____ ____

   b. Write this number in expanded notation.

   _____

6. Write the following expanded notation in standard form.
   $600 + 50 + 4 + 0.2 + 0.07 + 0.009$ _____

7. a. Build an 8-digit number. Use these clues.

   The digit in the place with the greatest value is equal to $4 + 0$.
   The digit in the place with the least value is equal to 3^2.
   The number in the hundreds place is the first counting number.
   The number in the tenths place multiplied by 54 is zero.
   The number in the tens place is the square root of 9.
   The number in the ones place is the square root of 4.
   The number in the hundredths place is the product of the number in the tens place
      and the number in the ones place.
   The number in the thousands place is equal to $9 - 2^2$.

   ___ ___ , ___ ___ ___ . ___ ___ ___

   b. Write this number in expanded notation.

   _____

LESSON 3·9   Math Boxes

1. Write five names for 1,000,000.

_____

_____

_____

_____

_____

_____

SRB 219

2. Use a straightedge to draw an angle that is less than 90°.

SRB 139

3. Write < or >.

3.67 _____ 3.7

0.02 _____ 0.21

4.06 _____ 4.02

3.1 _____ 3.15

7.6 _____ 7.56

SRB 9 32 33

4. I have four sides. All opposite sides are parallel. I have no right angles. Draw me in the space below.

I am called a _____.

SRB 143

5. What is the measure of angle R?

Q

27°

20° S

R

measure angle R = _____

SRB 207

6. Solve.

_____ = 3,000 * 800

_____ * 60 = 54,000

_____ = 40 * 900

20 * 5,000 = _____

72,000 = _____ * 900

SRB 18

LESSON 3·10 — The Geometry Template

Math Message

Answer the following questions about your Geometry Template. DO NOT count the protractors, Percent Circle, and little holes next to the rulers.

1. How many shapes are on the Geometry Template? _____

2. What fraction of these shapes are polygons? _____

3. What fraction of the shapes are quadrangles? _____

Problems for the Geometry Template

The problems on journal pages 93 and 95 are labeled Easy and Moderate. Each problem has been assigned a number of points according to its difficulty.

Complete as many of these problems as you can. Your Geometry Template and a sharp pencil are the only tools you may use. Record and label your answers on the page opposite the problems.

Some of the problems might seem confusing at first. Before asking your teacher for help, try the following:

◆ Look at the examples on the journal page. Do they help you understand what the problem is asking you to do?

◆ If you are not sure what a word means, look it up in the Glossary in your *Student Reference Book.* You might also look for help in the geometry section of the *Student Reference Book.*

◆ Find a classmate who is working on the same problem. Can the two of you work together to find a solution?

◆ Find a classmate who has completed the problem. Can she or he give you hints about how to solve it?

When the time for this activity has ended, total the number of points that you have scored. If you didn't have time to complete all these pages, you can continue working on them when you have free time.

Good luck and have fun!

LESSON 3·10 Problems for the Geometry Template *cont.*

Record your solutions on the next page. Include the problem numbers.

Easy

Examples

1. Using only shapes on your Geometry Template, draw an interesting picture. (2 points)

2. Trace all of the polygons on the Geometry Template that have at least one pair of **parallel sides.** (1 point each)

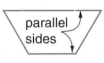

parallel sides

3. Trace all of the polygons on the Geometry Template that have no pairs of parallel sides. (1 point each)

4. Trace three polygons that have *at least* one **right angle** each, three polygons that have *at least* one **acute angle** each, and three polygons that have *at least* one **obtuse angle** each. ($\frac{1}{2}$ point each)

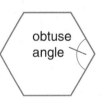

obtuse angle

5. Assume that the side of the largest square on the template has a length of 1 **unit.** Draw three different polygons, each with a **perimeter** of 8 units. (2 points each)

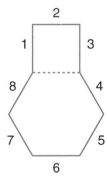

93

LESSON 3·10   Problems for the Geometry Template *cont.*

Solutions

LESSON 3·10 **Problems for the Geometry Template** *cont.*

Record your solutions on the next page. Include the problem numbers.

Moderate

Examples

6. Use your template to **copy** this design. (3 points)

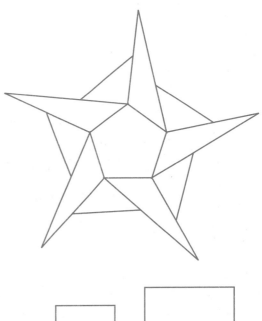

7. Without using a ruler, enlarge the rectangle.
   First draw a rectangle that is twice the size
   of the rectangle on the Geometry Template.
   Then draw a rectangle 3 times the size of the
   rectangle on the Geometry Template.
   (3 points each)

8. Compare the **perimeters** of the rectangle and the
   pentagon on the Geometry Template. Which
   polygon has the greater perimeter? You may not use
   the rulers on the template to help you. Describe how
   you were able to use other parts of your Geometry
   Template to solve this problem. (6 points)

9. Use the triangles on the Geometry Template to
   draw four different **parallelograms.** (2 points each)

10. Using any two polygons from the Geometry
    Template, draw five different **pentagons.**
    (2 points each)

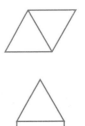

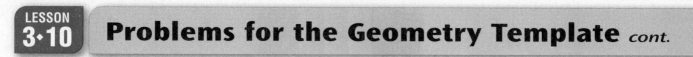

LESSON 3·10 **Problems for the Geometry Template** *cont.*

Solutions

LESSON 3·10

Math Boxes

1. What kind of regular polygon could be partially hidden behind the wall?

   _____

   Complete the shape.

   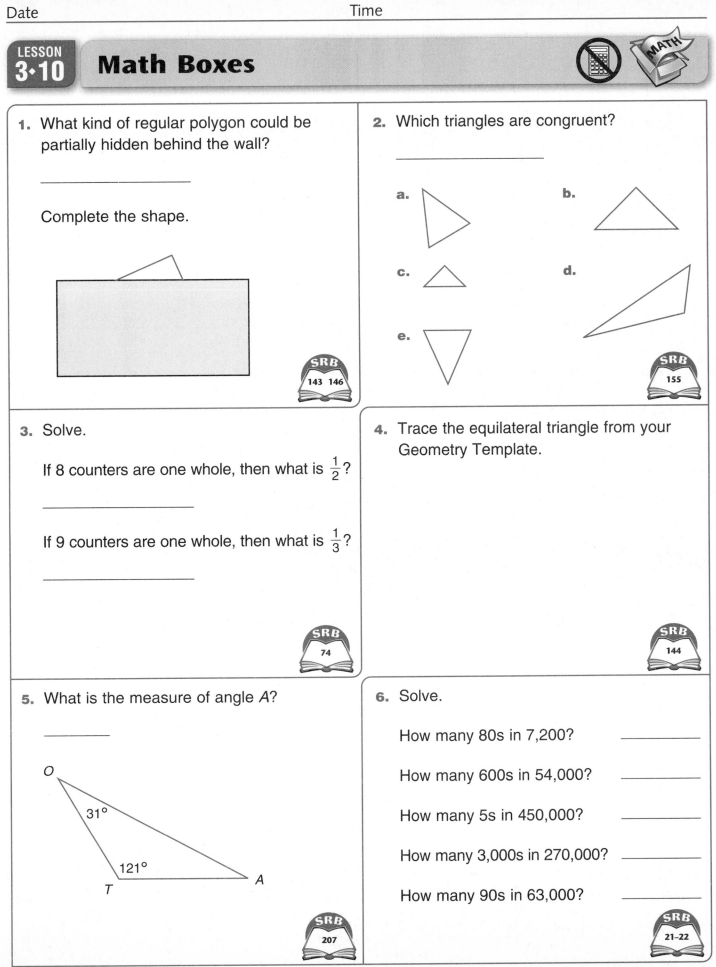

   SRB 143 146

2. Which triangles are congruent?

   _____

   a.

   b.

   c.

   d.

   e.

   SRB 155

3. Solve.

   If 8 counters are one whole, then what is $\frac{1}{2}$?

   _____

   If 9 counters are one whole, then what is $\frac{1}{3}$?

   _____

   SRB 74

4. Trace the equilateral triangle from your Geometry Template.

   SRB 144

5. What is the measure of angle A?

   _____

   O

   31°

   121°

   T A

   SRB 207

6. Solve.

   How many 80s in 7,200?   _____

   How many 600s in 54,000?   _____

   How many 5s in 450,000?   _____

   How many 3,000s in 270,000?   _____

   How many 90s in 63,000?   _____

   SRB 21–22

LESSON 3·11 Math Boxes

1. True or false? Write T or F.

4,908 is divisible by 3. _____

58,462 is divisible by 2. _____

63,279 is divisible by 9. _____

27,350 is divisible by 5. _____

77,922 is divisible by 6. _____

SRB 11

2. Solve.

$8 * 700 =$ _____

$36,000 =$ _____ $* 40$

$320,000 = 800 *$ _____

$2,000 *$ _____ $= 24,000$

$5,000 * 4,000 =$ _____

SRB 18 219

3. Write the prime factorization for 68.

_____

SRB 12

4. Solve.

If two counters are $\frac{1}{4}$, then
what is one whole? _____

If 15 counters are one whole,
then what is $\frac{1}{3}$? _____

SRB 74

5. Estimate and solve.

214
+182

Estimate: _____

Solution: _____

1,532
− 176

Estimate: _____

Solution: _____

SRB 13–17

6. Circle the best estimate for each problem.
Then solve.

$52.2 * 39.7$

| 2,000 | 20,000 | 200,000 |

Solution: _____

$148.3 * 232.51$

| 3,000 | 30,000 | 300,000 |

Solution: _____

SRB 38–39 45

LESSON 4·1 Mental Division Strategy

Fact knowledge can help you find how many times a 1-digit number will divide any large number.

Example: Divide 56 by 7 mentally.

Think: *How many 7s in 56?*   Or think: *7 times what number equals 56?*

Continue: *Since 7 * 8 = 56, there must also be 8 [7s] in 56. So 56 divided by 7 equals 8.*

Knowing basic facts helps you break the larger number into two or more friendly numbers—numbers that are easy to divide by the 1-digit number.

Example: Divide 96 by 3 mentally.

Break 96 into friendly numbers. Here are two ways.

◆ 90 and 6. Ask yourself: *How many 3s in 90?* (30) *How many 3s in 6?* (2)

   Total: 30 + 2 = 32

◆ 60 and 36. Ask yourself: *How many 3s in 60?* (20) *How many 3s in 36?* (12)

   Total: 20 + 12 = 32

So 96 divided by 3 equals 32. Check the result: 3 * 32 = 96.

Complete the following statements. List the friendly parts that you used.

1. 42 divided by 3 equals _____.

   _____
   (friendly parts for 42)

2. 68 divided by 4 equals _____.

   _____
   (friendly parts for 68)

3. 83 divided by 6 equals _____.

   _____
   (friendly parts for 83)

4. 99 divided by 7 equals _____.

   _____
   (friendly parts for 99)

5. Fifteen-year-old oak trees are often about 25 feet tall. Rose, a 15-year-old girl, is about 5 feet tall. How many times taller are the trees than Rose?

   _____

6. The job of interviewing 500 students in a school is to be divided equally among 10 interviewers. How many students should each interviewer talk to?

   _____

LESSON 4·1 | **Math Boxes**

1. Write the value of each of the following digits in the numeral 34,089,750.

    a. 4 _____

    b. 8 _____

    c. 5 _____

    d. 9 _____

    e. 3 _____

    SRB 4

2. Write the following numbers in standard notation.

    a. $6^2 =$ _____

    b. $10^5 =$ _____

    c. $14^2 =$ _____

    d. $8^3 =$ _____

    e. $3^4 =$ _____

    SRB 34–36 243

3. Roger had saved $10.05 from his allowance. Then he bought a paint-by-numbers kit for $7.39. How much money does he have left?

    _____

    SRB 5–7

4. Javier has $5.00 to buy school supplies. He wants one pack of pencils for $1.38, a notebook for $2.74, and some writing paper for $1.29. If he has enough money, how much change will he get back?

    _____

    If not, how much more money does he need?

    _____

    SRB 243

5. Use your full-circle protractor to measure angle *CAT*.

    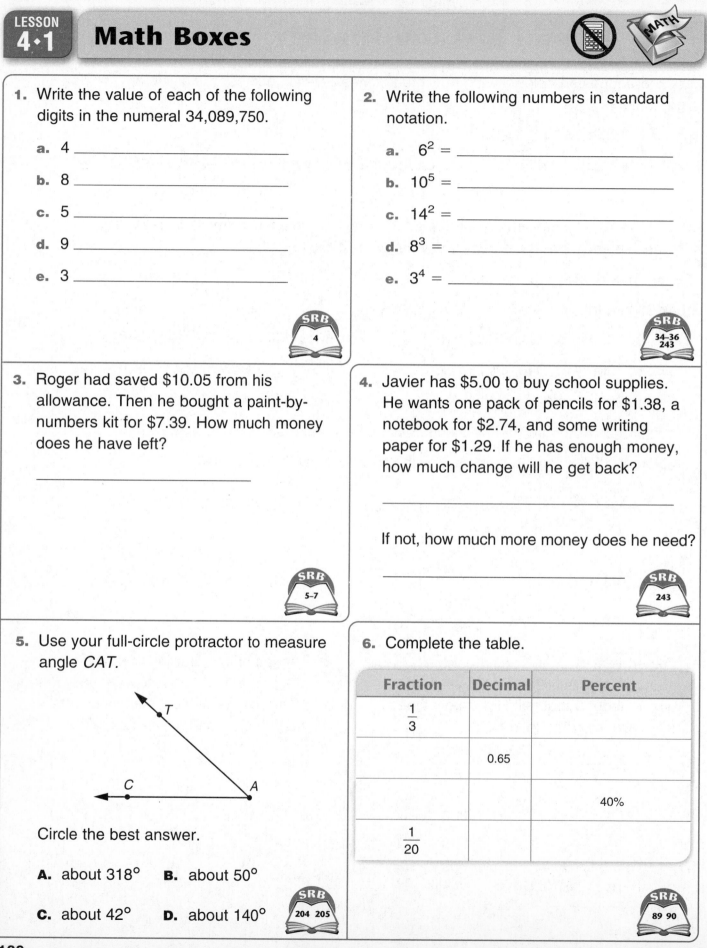

    Circle the best answer.

    A. about 318°   **B.** about 50°

    C. about 42°   **D.** about 140°

    SRB 204 205

6. Complete the table.

Fraction	Decimal	Percent
$\frac{1}{3}$		
	0.65	
		40%
$\frac{1}{20}$		

    SRB 89 90

LESSON 4·2 The Partial-Quotients Division Algorithm

Use the partial-quotients algorithm to solve these problems.

1. $6\overline{)495}$ _____

2. $832 \div 15 \rightarrow$ _____

3. $3,518 / 32 \rightarrow$ _____

4. $\dfrac{5,360}{54} \rightarrow$ _____

5. Jerry was sorting 389 marbles into bags. He put a
   dozen in each bag. How many bags does he need? _____

LESSON 4·2   **Math Boxes**

---

1. Write < or >.

   a. 0.45 _____ $\frac{3}{4}$

   b. 0.89 _____ $\frac{8}{10}$

   c. $\frac{4}{5}$ _____ 0.54

   d. $\frac{1}{3}$ _____ 0.35

   e. $\frac{7}{8}$ _____ 0.9

SRB
9 83
89

---

2. Sasha earns $4.50 per day on her paper route. She delivers papers every day. How much does she earn in two weeks?

Open sentence: _____

Solution: _____

Answer: _____

SRB
38–40
243

---

3. Write the prime factorization of 80.

_____

_____

SRB
12

---

4. Without using a protractor, find the measurement of the missing angle.

90°

50°   ____°

SRB
207

---

5. Solve.

   a. $\begin{array}{r} 209.0 \\ -73.5 \\ \hline \end{array}$ **b.** $\begin{array}{r} \$30.49 \\ -\$8.51 \\ \hline \end{array}$

   c. $\begin{array}{r} 4.339 \\ +6.671 \\ \hline \end{array}$ **d.** $\begin{array}{r} 25.03 \\ +14.58 \\ \hline \end{array}$

SRB
34–36

6.

The United States in 1900

40% Urban | 60% Rural

Circle the best answer.

A. In 1900, more than half of the communities were rural.

B. In 1900, 6 out of 10 communities in the United States were rural.

C. In 1900, more than $\frac{3}{4}$ of communities in the United States were rural.

SRB
125

LESSON 4·3

Distances between U.S. Cities

1. Write 2 questions about map scales that can be answered using page 211 of your *Student Reference Book*.

_____

_____

_____

_____

2. Use the map of the United States on pages 386 and 387 of your *Student Reference Book* to estimate the distances between the following cities. Measure each map distance in inches. Complete the table.
(Scale: 1 inch represents 200 miles)

Cities	Map Distance (inches)	Real Distance (miles)
Chicago, IL, to Pittsburgh, PA	2 in.	400 mi
Little Rock, AR, to Jackson, MS		
San Francisco, CA, to Salt Lake City, UT		
Indianapolis, IN, to Raleigh, NC		
Chicago, IL, to Boston, MA		
San Antonio, TX, to Buffalo, NY		
Salt Lake City, UT, to Pierre, SD		

3. Explain how you found the real distance from Salt Lake City, UT, to Pierre, SD.

_____

_____

_____

_____

LESSON 4·3   Finding Factors

Example 1:

$$\overset{\frown}{1 \quad \overset{\frown}{2 \quad \overset{\frown}{4 \quad 8} \quad} 16}$$

1. Make a factor rainbow to list all the factors of the number 36.

2. **a.** Fill in the blanks in the table.

Product	Exponential Notation	Square Number
		2,500

   b. The square root of 2,500 is _____.

3. **a.** Find factor strings for the number 52.

   b. The prime factorization for 52 is

   _____.

Number	Factor Strings	Length
52		

   c. $52 \div 13 =$ _____

Example 2:

$$\begin{array}{c} 12 \\ / \ \backslash \\ 3 \ * \ 4 \\ / \quad / \backslash \\ 3 \ * \ 2 \ * \ 2 \end{array}$$

4. **a.** Make a factor tree to find the prime factorization for the number 72.

   b. $2^3 * 3^2 =$ _____

5. **a.** Use the divisibility rules for 1, 2, 3, 4, 5, 6, 9, and 10 to find factor pairs for 80.

   _____

   b. $16\overline{)80}$ _____

6. Find all the factors of the number 54. Use the method of your choice.

   _____

LESSON 4·3

Math Boxes

1. Write the value of each digit in the numeral 4,231.756.

   a. 5 _____

   b. 7 _____

   c. 3 _____

   d. 2 _____

   e. 6 _____

   SRB 28–31

2. Larry spent $4.82 on a notebook, $1.79 on paper to fill it, and $2.14 on a pen. How much did he spend in all? Fill in the circle next to the best answer.

   Ⓐ $7.75

   Ⓑ $8.75

   Ⓒ $8.65

   Ⓓ $7.65

   SRB 34–36 243

3. Use a calculator to rename each of the following in standard notation.

   a. $24^2 =$ _____

   b. $11^3 =$ _____

   c. $9^4 =$ _____

   d. $4^5 =$ _____

   e. $2^7 =$ _____

   SRB 5–7

4. Use your full circle protractor to measure angle A.

   $\angle A$ is about _____ °.

   SRB 204 205

5. Gustavo got his driver's license in the year 2004 when he was 16 years old. In what year was he born?

   Open sentence:

   _____

   Solution: _____

   Answer: _____

   SRB 243 247

6. Complete the table.

Fraction	Decimal	Percent
		95%
	0.80	
$\frac{3}{9}$		
$\frac{6}{8}$		
		$66\frac{2}{3}\%$

   SRB 89 90

105

LESSON 4·4 The Partial-Quotients Algorithm

Example: 185 / 8 → ?

One way:	**Another way:**	**Another way:**

One way:

```
8)185 |
 -80  | 10
 ————
 105  |
 -80  | 10
 ————
  25  |
 -24  | 3
 ————
   1  | 23
```

Another way:

```
8)185  |
-160   | 20
————
  25   |
 -24   | 3
————
   1   | 23
```

Another way:

```
8)185
```
Rename 185 using
multiples of 8:
160 + 24 + 1
Think: 160 = 20 [8s]
24 = 3 [8s]
20 + 3 = 23 [8s] with
1 left over

The answer, 23 R1, is the same for each way.

Use the partial-quotients algorithm to solve these problems.

1. 64 ÷ 8 = _____

2. 749 / 7 = _____

3. 2,628 ÷ 36 = _____

4. 8,190 / 9 = _____

5. Raoul has 237 string bean seeds. He plants them in rows with 8 seeds in each row. How many complete rows can he plant?

Estimate: _____

Solution: _____ rows

LESSON 4·4

The Partial-Quotients Algorithm *continued*

Divide.

6. 823 / 3 → _____

7. 2,815 ÷ 43 → _____

8. 4,290 / 64 → _____

9. Regina put 1,610 math books into boxes.
    Each box held 24 books. How many boxes did she use?

    Estimate: _____

    Solution: _____ boxes

10. Make up a number story that can be solved with division.
    Solve it using a division algorithm.

    _____

    _____

    _____

    _____

    Solution: _____

LESSON 4·4 — Math Boxes

1. Write $<$ or $>$.

a. $\dfrac{3}{5}$ _____ 0.70

b. $\dfrac{1}{4}$ _____ 0.21

c. 0.38 _____ $\dfrac{3}{10}$

d. 0.6 _____ $\dfrac{2}{3}$

e. 0.95 _____ $\dfrac{90}{100}$

SRB
9 83
89

2. Jamie bikes 18.5 mi per day. How many miles will she ride in 13 days?

Open sentence: _____

Solution: _____

Answer: _____

SRB
38–40
243

3. Write the prime factorization of 132.

_____

_____

SRB
12

4. Without using a protractor, find the measurement of the missing angle.

79°

120°

102°

o

SRB
207

5. Solve.

a. $2.03 - 0.76 =$ _____

b. _____ $= 57.97 + 3.03$

c. _____ $= 691.23 + 507.26$

d. $29.05 + 103.94 =$ _____

SRB
34–36

6. Fill in the circle next to the best answer.

Favorite 5th Grade Colors

blue

red

yellow green

○ **A.** More than $\dfrac{1}{2}$ of the students chose blue.

○ **B.** 50% of the students chose yellow or green.

○ **C.** More than 25% of the students chose yellow or red.

SRB
125

LESSON 4·5   # Estimate and Calculate Quotients

For each problem:

◆ Make a magnitude estimate of the quotient. Ask yourself: *Is the answer in the tenths, ones, tens, or hundreds?*

◆ Circle a box to show the magnitude of your estimate.

◆ Write a number sentence to show how you estimated.

◆ If there is a decimal point, divide as if the numbers were whole numbers.

◆ Use your magnitude estimate to place the decimal point in the final answer.

◆ Check that your final answer is reasonable.

1. $3\overline{)36.6}$

0.1s	1s	10s	100s

How I estimated: _____

Answer: _____

2. $4\overline{)9.48}$

0.1s	1s	10s	100s

How I estimated: _____

Answer: _____

3. $18.55 \div 7$

0.1s	1s	10s	100s

How I estimated: _____

Answer: _____

4. $7.842 \div 6$

0.1s	1s	10s	100s

How I estimated: _____

Answer: _____

5. $560.1 / 3$

0.1s	1s	10s	100s

How I estimated: _____

Answer: _____

6. $3.84 / 6$

0.1s	1s	10s	100s

How I estimated: _____

Answer: _____

LESSON 4·5

Math Boxes

1.

Population of Maine	
Year	**Population**
1900	694,000
1950	914,000
2000	1,259,000

Predict the population for 2010.
Circle the best answer.

A. 950,000   B. 1,400,000

C. 2,000,000   D. 2,400,000

SRB 230

2. Measure each line segment to the nearest centimeter.

_____

_____ cm

_____

_____ cm

SRB 183

3. Use a straightedge to draw an obtuse angle. Label it $\angle T$.

Estimate $m\angle T$: _____

$m\angle T =$ _____

SRB 138 139 204 205

4. Solve.

a. $10 \times 0.1 =$ _____

b. $0.7 \times 10 =$ _____

c. $18 \times 0.1 =$ _____

d. $10 * 40 =$ _____

e. $305 \times 0.1 =$ _____

SRB 18 37

5. Fill in the missing numbers.

a. $(2.5 + 6) + 3 = 2.5 + ($_____$ + 3)$

b. $73.426 +$ _____ $= 1.39 + 73.426$

c. $(4 * 5) \times 2 = 4 \times (5 *$ _____$)$

d. $3.6 \times$ _____ $= 1.435 \times 3.6$

SRB 224

6. Circle all the fractions that are equivalent to $\frac{9}{18}$.

$\frac{7}{14}$ $\frac{7}{8}$ $\frac{6}{9}$ $\frac{5}{10}$ $\frac{2}{3}$

SRB 59–61

LESSON 4·6 Interpreting Remainders

For each number story:

◆ Draw a picture, and write an open sentence.

◆ Use a division algorithm to solve the problem.

◆ Tell what the remainder represents.

◆ Decide what to do about the remainder.

1. Compact discs are on sale for $9, including tax. How many can you buy with $30?

Picture:

Number sentence: _____

_____

Answer: _____ compact discs

What does the remainder represent?

_____

_____

Circle what you did about the remainder.

Ignored it

Reported it as a fraction or decimal

Rounded the answer up

2. Rebecca and her three sisters bought their mother a bread machine for her birthday. The machine cost $219, including tax. The sisters split the bill evenly. How much did each sister contribute?

Picture:

Number sentence: _____

_____

Answer: $_____

What does the remainder represent?

_____

_____

Circle what you did about the remainder.

Ignored it

Reported it as a fraction or decimal

Rounded the answer up

LESSON 4·6 **Interpreting Remainders** *continued*

3. You are organizing a trip to a museum for 110 students, teachers, and parents. If each bus can seat 25 people, how many buses do you need?

Picture:

Number sentence: _____

_____

Solution: _____ buses

What does the remainder represent?

_____

_____

_____

Circle what you did about the remainder.

Ignored it

Reported it as a fraction or decimal

Rounded the answer up

Review: Magnitude Estimates and Division

4. $15\overline{)4,380}$

| 0.1s | 1s | 10s | 100s |

How I estimated: _____

Answer: _____

5. $3\overline{)70.5}$

| 0.1s | 1s | 10s | 100s |

How I estimated: _____

Answer: _____

6. 82.8 / 12

| 0.1s | 1s | 10s | 100s |

How I estimated: _____

Answer: _____

Try This

7. 3.75 / 25

| 0.1s | 1s | 10s | 100s |

How I estimated: _____

Answer: _____

LESSON 4·6   Place-Value Puzzles

1. The digit in the thousands place is 6.

    The digit in the ones place is the sum of the digits in a dozen.

    The digit in the millions place is $\frac{1}{10}$ of 70.

    The digit in the hundred-thousands place is $\frac{1}{2}$ of the digit in the thousands place.

    The digit in the hundreds place is the sum of the digit in the thousands place and the digit in the ones place.

    The rest of the digits are all 5s.   ___ ___ , ___ ___ ___ , ___ ___ ___

2. The digit in the tens place is 2.

    The digit in the ones place is double the digit in the tens place.

    The digit in the hundreds place is three times the digit in the tens place.

    The digit in the hundred-thousands place is an odd number less than 3.

    The digit in the millions place is $\frac{1}{3}$ of 15.

    The rest of the digits are all 9s.   ___ ___ , ___ ___ ___ , ___ ___ ___

3. The digit in the ten-thousands place is the sum of the digits in 150.

    The digit in the millions place is a prime number greater than 5.

    The digit in the hundreds place is $\frac{1}{2}$ of the digit in the thousands place.

    The digit in the tenths place is 1 less than the digit in the millions place.

    The digit in the thousands place is $\frac{2}{5}$ of 20.

    The rest of the digits are all 3s.   ___ , ___ ___ ___ , ___ ___ ___ . ___ ___

Try This

4. The digit in the thousands place is the smallest square number greater than 1.

    The digit in the tens place is the same as the digit in the place 1,000 times greater.

    The digit in the ten-thousands place is $\frac{1}{2}$ of the digit in the ten-millions place.

    The digit in the ten-millions place is two more than the digit in the thousands place.

    The digit in the hundreds place is 1 greater than double the digit in the ten-thousands place.

    The rest of the digits are all 2s.   ___ ___ , ___ ___ ___ , ___ ___ ___

LESSON 4·6 Math Boxes

1. Write < or >.

a. $\frac{1}{3}$ _____ 0.5

b. 0.25 _____ $\frac{1}{5}$

c. 0.85 _____ $\frac{2}{5}$

d. $\frac{1}{3}$ _____ 0.4

e. 0.5 _____ $\frac{6}{10}$

SRB 9 83 89

2. Isabel is making 12 bows for a special project. She needs 15.25 in. of ribbon for each bow. How much ribbon does she need all together?

Open sentence: _____

Solution: _____

Answer: _____

SRB 243 247

3. Write the prime factorization for 200.

_____

SRB 12

4. Without using a protractor, find the measurement of the missing angle.

60°   45°

SRB 207

5. Solve.

a. 14.59 + 202.7 = _____

b. 89 + 36.02 = _____

c. _____ = 60.07 − 0.08

d. _____ = 15.76 − 5.99

SRB 34–36

6. Fill in the circle next to the best answer.

Ⓐ In 2000, about $\frac{3}{4}$ of the people in the United States lived in small communities.

The United States in 2000

23% Rural

77% Urban

Ⓑ In 2000, more than $\frac{1}{4}$ of the people in the United States lived in rural areas.

Ⓒ In 2000, about $\frac{3}{4}$ of the people in the United States lived in urban communities.

SRB 125

LESSON 4·7

Calculator Exploration: Division

Stored Operations:

Record the key sequences for your stored operations.

```
┌─────────────────────────────────────────────────────────────────────┐
│                                                                     │
│                                                                     │
│                                                                     │
│                                                                     │
│                                                                     │
└─────────────────────────────────────────────────────────────────────┘
```

Write 3-, 4-, and 5-digit numbers from the Class Data Pad. Make a magnitude estimate for each number squared. Then use your calculator to square each number and record the result.

10,000s	100,000s	1,000,000s	10,000,000s	100,000,000s	1,000,000,000s

1. _____ $\boxed{\wedge}$ 2 = _____

10,000s	100,000s	1,000,000s	10,000,000s	100,000,000s	1,000,000,000s

2. _____ $\boxed{\wedge}$ 2 = _____

10,000s	100,000s	1,000,000s	10,000,000s	100,000,000s	1,000,000,000s

3. _____ $\boxed{\wedge}$ 2 = _____

10,000s	100,000s	1,000,000s	10,000,000s	100,000,000s	1,000,000,000s

4. _____ $\boxed{\wedge}$ 2 = _____

10,000s	100,000s	1,000,000s	10,000,000s	100,000,000s	1,000,000,000s

5. _____ $\boxed{\wedge}$ 2 = _____

6. What observations can you make about the magnitude of 3-, 4-, and 5-digit numbers squared?

_____

_____

115

LESSON 4·7 Triangle and Polygon Review

Fill in the oval next to the correct answer for each triangle.

1.    **2.**    **3.**    **4.**    **5.**

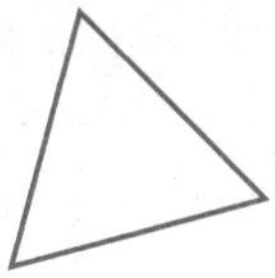

○ equilateral	○ equilateral	○ equilateral	○ equilateral	○ equilateral
○ isosceles	○ isosceles	○ isosceles	○ isosceles	○ right
○ scalene	○ right	○ right	○ scalene	○ scalene

6. Marlene drew four shapes—an isosceles triangle, a pentagon, a trapezoid, and a rectangle. She covered up most of each figure as shown below. Write the name below each figure. Draw the rest of the figure.

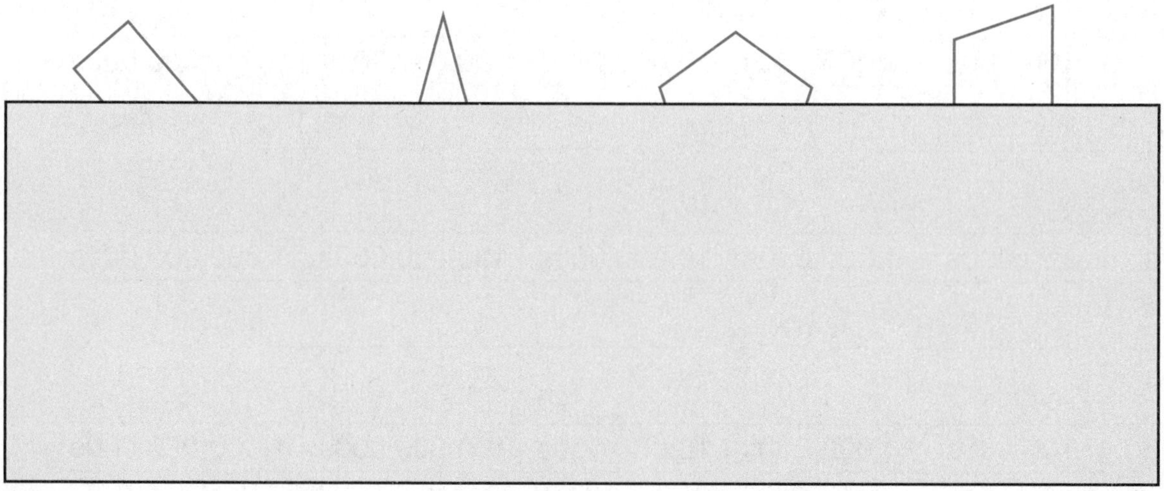

Try This

7. What is the measure of each angle in an equilateral triangle? _____

Explain how you know. _____

_____

_____

Math Boxes

1. Predict the number of words used at age 3. Fill in the circle next to the best answer.

Age in Years	Number of Words
$2\frac{1}{2}$	446
$3\frac{1}{2}$	1,222

Ⓐ 515   Ⓑ 902

Ⓒ 1,540   Ⓓ 1,870

SRB
121

2. Measure each line segment to the nearest quarter-inch.

_____

_____ in.

_____

_____ in.

SRB
183

3. Use a straightedge to draw an acute angle. Label it ∠A.

a. Estimate m∠A: _____

b. m∠A = _____

SRB
138

4. Solve.

a. $8 * 10 =$ _____

b. $600 * 0.1 =$ _____

c. $0.79 * 10 =$ _____

d. $900 * 0.1 =$ _____

e. $90.6 * 10 =$ _____

f. $234 * 0.1 =$ _____

SRB
38–40

5. Fill in the missing numbers.

a. $3,624 + 72,603 = 72,603 +$ _____

b. $942,136 *$ _____ $= 74.05 * 942,136$

c. $5.6 + (0.17 + 126) =$
$(5.6$ _____ $) + 126$

d. $69 * ($ _____ $* 426) = (69 * 426) * 1.4$

e. _____ $* 0.167 = 0.167 * 0.6$

6. Circle all the fractions that are equivalent to $\frac{4}{12}$.

$\frac{5}{15}$ $\frac{2}{6}$ $\frac{8}{16}$ $\frac{3}{9}$ $\frac{12}{16}$

SRB
59–61

LESSON 4·7 *Algebra Election*

Materials

- [ ] 32 *First to 100* Problem Cards
  (*Math Masters,* pp. 456 and 457)
- [ ] Electoral Vote Map
  (*Math Masters,* pp. 442 and 443)
- [ ] 1 six-sided die
- [ ] 4 pennies or other small counters
- [ ] calculator

Players

2 teams, each with 2 players

Object of the game

Players move their counters on a map of the United States. For each state, or the District of Columbia (D.C.), that a player lands on, the player tries to win that state's electoral votes by solving a problem. The first team to collect 270 or more votes wins the election. Winning-team members become President and Vice President.

Directions

1. Each player puts a counter on Iowa.
2. One member of each team rolls the die. The team with the higher roll goes first.
3. Alternate turns between teams and partners: Team 1, Player 1; Team 2, Player 1; Team 1, Player 2; Team 2, Player 2.
4. Shuffle the Problem Cards. Place them facedown in a pile.
5. The first player rolls the die. The result tells how many moves the player must make from the current state. Each new state counts as one move. Moves can be in any direction as long as they pass between states that share a common border. *Exceptions:* Players can get to and from Alaska by way of Washington state, and to and from Hawaii by way of California. Once a player has been in a state, the player may not return to that state on the same turn.
6. The player makes the indicated number of moves and puts the counter on the last state moved to. The map names how many electoral votes the state has.

How many inches are there in x feet? How many centimeters are there in x meters? **1**	How many quarts are there in x gallons? **2**	What is the smallest number of x's you can add to get a sum greater than 100? **3**	Is $50 * x$ greater than 1,000? Is $\frac{x}{10}$ less than 1? **4**
$\frac{1}{2}$ of $x = ?$ $\frac{1}{10}$ of $x = ?$ **5**	$1 - x = ?$ $x + 998 = ?$ **6**	If x people share 1,000 stamps equally, how many stamps will each person get? **7**	What time will it be x minutes from now? What time was it x minutes ago? **8**
It is 102 miles to your destination. You have gone x miles. How many miles are left? **9**	What whole or mixed number equals x divided by 2? **10**	Is x a prime or a composite number? Is x divisible by 2? **11**	The time is 11:05 A.M. The train left x minutes ago. What time did the train leave? **12**
Bill was born in 1939. Freddy was born the same day, but x years later. In what year was Freddy born? **13**	Which is larger: $2 * x$ or $x + 50$? **14**	There are x rows of seats. There are 9 seats in each row. How many seats are there in all? **15**	Sargon spent x cents on apples. If she paid with a $5 bill, how much change should she get? **16**

The temperature was 25°F. It dropped x degrees. What is the new temperature? **17**	Each story in a building is 10 feet high. If the building has x stories, how tall is it? **18**	Which is larger: $2 * x$ or $\frac{100}{x}$? **19**	$20 * x = ?$ **20**
Name all of the whole-number factors of x. **21**	Is x an even or an odd number? Is x divisible by 9? **22**	Shalanda was born on a Tuesday. Linda was born x days later. On what day of the week was Linda born? **23**	Will had a quarter plus x cents. How much money did he have in all? **24**
Find the perimeter and area of this square. x cm □ x cm **25**	What is the median of these weights? 5 pounds 21 pounds x pounds What is the range? **26**	$x°$ $70°$ **27**	$x^2 = ?$ 50% of $x^2 = ?$ **28**
$(3x + 4) - 8 = ?$ **29**	x out of 100 students voted for Ruby. Is this more than 25%, less than 25%, or exactly 25% of the students? **30**	There are 200 students at Wilson School. x% speak Spanish. How many students speak Spanish? **31**	People answered a survey question either Yes or No. x% answered Yes. What percent answered No? **32**

LESSON 4·7

Algebra Election *continued*

7. The player takes the top Problem Card. The state's number of electoral votes is substituted for the variable *x* in the problem(s) on the card. The player solves the problem(s) and offers an answer. The other team checks the answer with a calculator.

8. If the answer is correct, the player's team wins the state's electoral votes. They do the following:

NOTE: Alaska and Hawaii are not drawn to scale.

   ◆ Write the state's name and its electoral votes on a piece of scratch paper.
   ◆ Write their first initials in pencil on the state to show that they have won it.
   Once a state is won, it is out of play. The opposing team may land on the state, but they cannot get its votes.

9. If the partners do not solve the problem(s) correctly, the state remains open. Players may still try to win its votes.

10. The next player rolls the die and moves his or her counter.

11. The first team to get at least 270 votes wins the election.

12. When all the Problem Cards have been used, shuffle the deck, and use it again.

13. Each player begins a turn from the last state he or she landed on.

Notes

◆ *A state* means *a state* or *the District of Columbia (D.C.)*.

◆ Partners may discuss the problem with one another. Each player, however, has to answer the problem on his or her own.

◆ If a player does not want to answer a Problem Card, the player may say "Pass" and draw another card. A player may pass 3 times during a game.

◆ If a Problem Card contains several problems, a player must answer all the questions on a card correctly to win a state's votes.

◆ Suggested strategy: Look at the map to see which states have the most votes, then work with your partner to win those states.

Variations

1. Agree on a time limit for answering problems.

2. Give one extra point if the player can name the capital of the state landed on.

3. A shorter version of the game can be played by going through all 32 cards just once. The team with the most votes at that time is the winner.

LESSON 4·8 Math Boxes

1. Complete the table.

Fraction	Decimal	Percent
$\frac{1}{5}$		
		38%
	0.75	
$\frac{4}{6}$		
		62.5%

SRB 89 90

2. Make up a set of at least 12 numbers that have the following landmarks.

minimum: 3

maximum: 9

median: 7

mode: 7

_____

_____

_____

SRB 119

3. Label these points on the number line.

0.9   0.56   0.25   0.1   0.7   0.4

0   0.5   1.0

SRB 32

4. Write 5 names for $\frac{1}{4}$.

_____

_____

_____

SRB 59–61

5. Tracy scored 95, 82, 90, and 83 on four tests. After the fifth test, the mode of her scores was 90. What did she score on the fifth test?

Score: _____

What was the mean of her 5 tests? _____

SRB 119–121

LESSON 5·1 Parts and Wholes

Work with a partner. Use counters to help you solve these problems.

1. This set has 15 counters. What fraction of the set is black?

    _____

2. If 12 counters are the whole set,
    what fraction of the set is 8 counters?

    _____

3. If 12 counters are the whole set,
    how many counters are $\frac{1}{4}$ of a set?

    _____ counters

4. If 20 counters are a whole,
    how many counters make $\frac{4}{5}$?

    _____ counters

5. If 6 counters are $\frac{1}{2}$ of a set,
    how big is the set?

    _____ counters

6. If 12 counters are $\frac{3}{4}$ of a set,
    how many counters are in the whole set?

    _____ counters

7. If 8 counters are a whole set, how many
    counters are in one and one-half sets?

    _____ counters

8. If 6 counters are two-thirds of a set, how many
    counters are in one and two-thirds sets?

    _____ counters

LESSON 5·1 — Finding Fractions of a Whole

1. In a school election, 141 fifth graders voted. One-third voted for Shira and two-thirds voted for Bree.

   141 votes

   $\frac{1}{3}$ Shira $\frac{2}{3}$ Bree

   a. How many votes did Shira get? _____

   b. How many votes did Bree get? _____

2. Bob, Liz, and Eli drove from Chicago to Denver.

   Bob drove $\frac{1}{10}$ of the distance.

   Liz drove $\frac{4}{10}$ of the distance.

   Eli drove $\frac{1}{2}$ of the distance.

   How many miles did each person drive?

   Chicago
   1,050 miles
   IL
   IA
   CO | NE
   Denver

   a. Bob: _____ miles   b. Liz: _____ miles   c. Eli: _____ miles

   Check to make sure that the total is 1,050 miles.

3. Carlos and Rick paid $8.75 for a present. Carlos paid $\frac{2}{5}$ of the total amount, and Rick paid $\frac{3}{5}$ of the total.

   a. How much did Carlos pay? _____

   b. How much did Rick pay? _____

4. A pizza costs $12.00, including tax. Scott paid $\frac{1}{4}$ of the total cost. Trung paid $\frac{1}{3}$ of the total cost. Iesha paid $\frac{1}{6}$. Bill paid the rest. How much did each person pay?

   a. Scott: $_____   b. Trung: $_____   c. Iesha: $_____   d. Bill: $_____

5. If 60 counters are the whole, how many counters make two-thirds? _____ counters

6. If 75 counters are $\frac{3}{4}$ of a set, how many counters are in the whole set? _____ counters

7. If 15 counters are a whole, how many counters make three-fifths? _____ counters

LESSON 5·1 | **Math Boxes**

---

1. Write a 10-digit numeral that has

   7 in the billions place,
   5 in the hundred-thousands place,
   3 in the ten-millions place,
   4 in the tens place,
   8 in the hundreds place, and
   2 in all other places.

   ___, ___ ___ ___, ___ ___ ___, ___ ___ ___

   Write the numeral in words.

   _____

   _____

   _____

   _____

   SRB 4

2. Write each fraction as a whole number or a mixed number.

   a. $\frac{24}{8} =$ _____

   b. $\frac{18}{5} =$ _____

   c. $\frac{21}{6} =$ _____

   d. $\frac{15}{4} =$ _____

   e. $\frac{11}{3} =$ _____

   SRB 23 62

---

3. Sixty students voted for their favorite fruit. The circle graph shows the results.

   a. What fraction voted for apples?

   _____

   b. What fraction voted for peaches?

   _____

   c. What fraction voted for strawberries?

   _____

   Favorite Fruits

   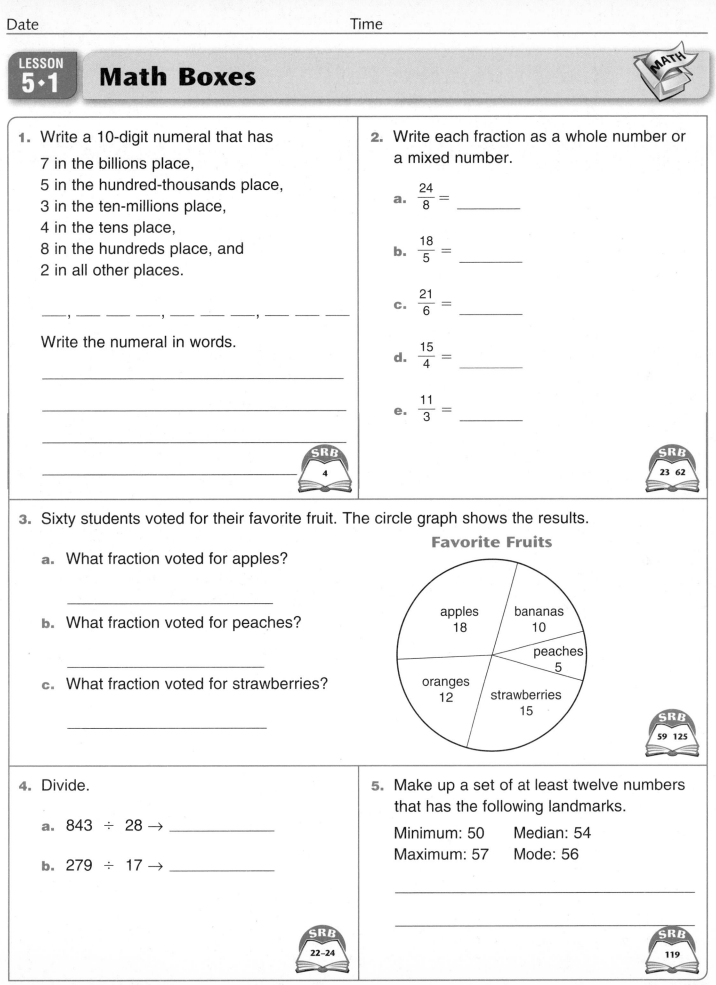

   apples 18
   bananas 10
   peaches 5
   oranges 12
   strawberries 15

   SRB 59 125

---

4. Divide.

   a. 843 ÷ 28 → _____

   b. 279 ÷ 17 → _____

   SRB 22–24

5. Make up a set of at least twelve numbers that has the following landmarks.

   Minimum: 50   Median: 54
   Maximum: 57   Mode: 56

   _____

   _____

   SRB 119

LESSON 5·2 Mixed Numbers: Part 1

Fractions greater than 1 can be written in several different ways.

Example: If ◯ is worth 1, what is ◯ ◯ ◔ worth?

The mixed-number name is $2\frac{3}{4}$. ($2\frac{3}{4}$ means $2 + \frac{3}{4}$.)

The fraction name is $\frac{11}{4}$. (*Think quarters:* ⊕ ⊕ ◔.)

So $2\frac{3}{4}$, $2 + \frac{3}{4}$, and $\frac{11}{4}$ are just different names for the same number.

In the problems below, the hexagon shape is worth 1.

Rule

hexagon

1. ⬡ = _____

2. △ = _____

3. ▱ = _____

4. ▱ = _____

Rule

hexagon

In the problems below, the hexagon shape is worth 1.
Write the mixed-number name and the fraction name
shown by each diagram.

5. ⬡ ▷   Mixed number = _____   Fraction = _____

6. ⬡ ⬢   Mixed number = _____   Fraction = _____

7. ⬡ ⬡ ▱   Mixed number = _____   Fraction = _____

8. ⬡ ⊏⊐⫿   Mixed number = _____   Fraction = _____

9. ⬡ ⬡ ⬡ ◔   Mixed number = _____   Fraction = _____

LESSON 5·2   **Mixed Numbers: Part 2**

For Problems 1–5, each triangle block is worth $\frac{1}{4}$.

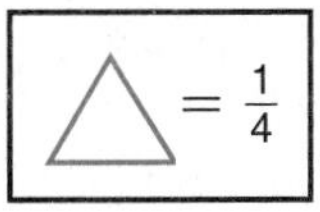

$\triangle = \frac{1}{4}$

Use your 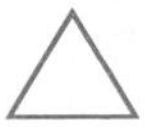  △, ▱, and ⬠ pattern blocks to solve these problems.

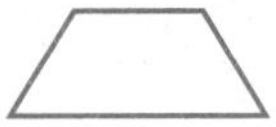

1. Cover a rhombus block with triangle blocks. A rhombus is worth _____.

2. Cover a trapezoid block with triangle blocks. A trapezoid is worth _____.

3. Arrange your blocks to make a shape worth 1.
   Trace the outline of each block that is part of
   your shape, or use your Geometry Template.
   Label each part with a fraction.

4. Arrange your blocks to make a shape that is worth $2\frac{1}{2}$. Trace the outline of each
   block that is part of your shape, or use your Geometry Template. Label each part
   with a fraction.

5. Use your blocks
   to cover this shape.

   Trace the outline of each
   block and label each part
   with a fraction.

   How much is the shape worth?

   _____

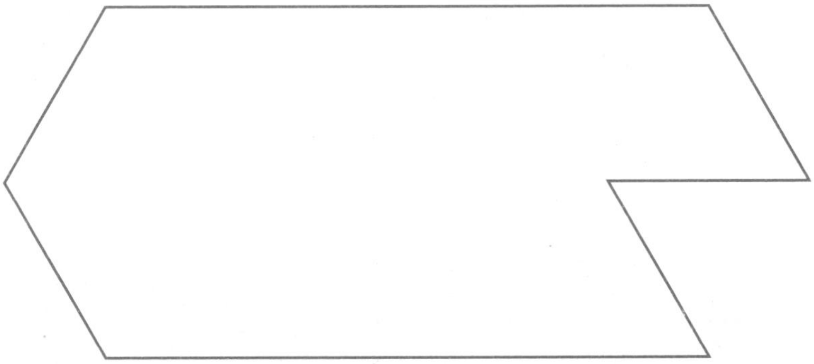

LESSON 5·2   **Mixed Numbers: Part 2** *continued*

For Problems 6–10, each triangle block is worth $\frac{1}{2}$.

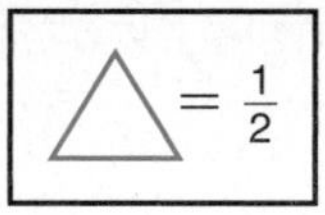

$\triangle = \frac{1}{2}$

Use your  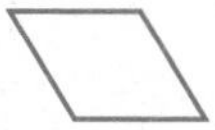 $\triangle$, $\diagup\!\!\!\diagdown$, and $\diagup\overline{}\diagdown$ pattern blocks to solve these problems.

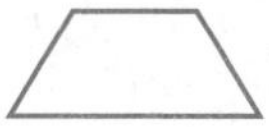

6. What shape is worth ONE? _____

7. A rhombus is worth _____.

8. A trapezoid is worth _____.

9. Arrange your blocks to make a shape that is worth $3\frac{1}{2}$. Trace the outline of each block that is part of your shape, or use your Geometry Template. Label each part with a fraction.

10. Use your blocks to cover the shape below. Trace the outline of each block. Label each part with a fraction.

    How much is the shape worth?

    _____

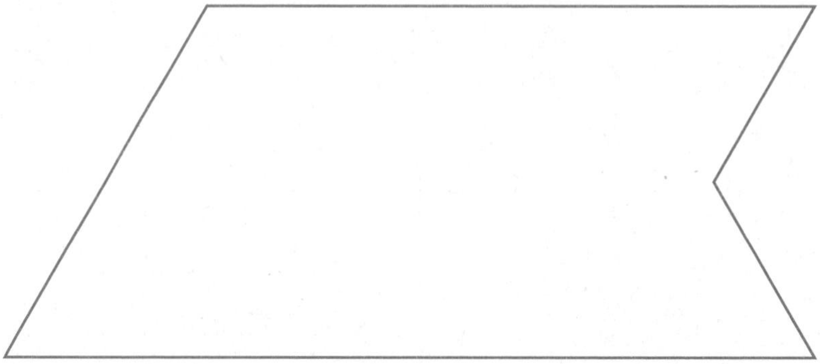

11. If a triangle block is $\frac{1}{4}$, make a diagram to show the fraction $\frac{15}{4}$.

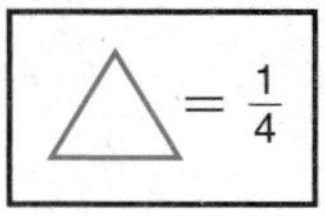

$\triangle = \frac{1}{4}$

Write $\frac{15}{4}$ as a mixed number. $\frac{15}{4} =$ _____.

LESSON 5·2 **Fractions on a Ruler**

1. Find and mark each of these lengths on the ruler below. Write the letter above the mark. Letters A and B are done for you.

   A: 5"   B: $\frac{1}{2}$"   C: $3\frac{1}{2}$"   D: $2\frac{1}{2}$"

   E: $4\frac{3}{4}$"   F: $\frac{1}{4}$"   G: $4\frac{1}{8}$"   H: $1\frac{7}{8}$"

   I: $1\frac{3}{8}$"   J: $\frac{15}{16}$"   K: $3\frac{1}{16}$"   L: $5\frac{9}{16}$"

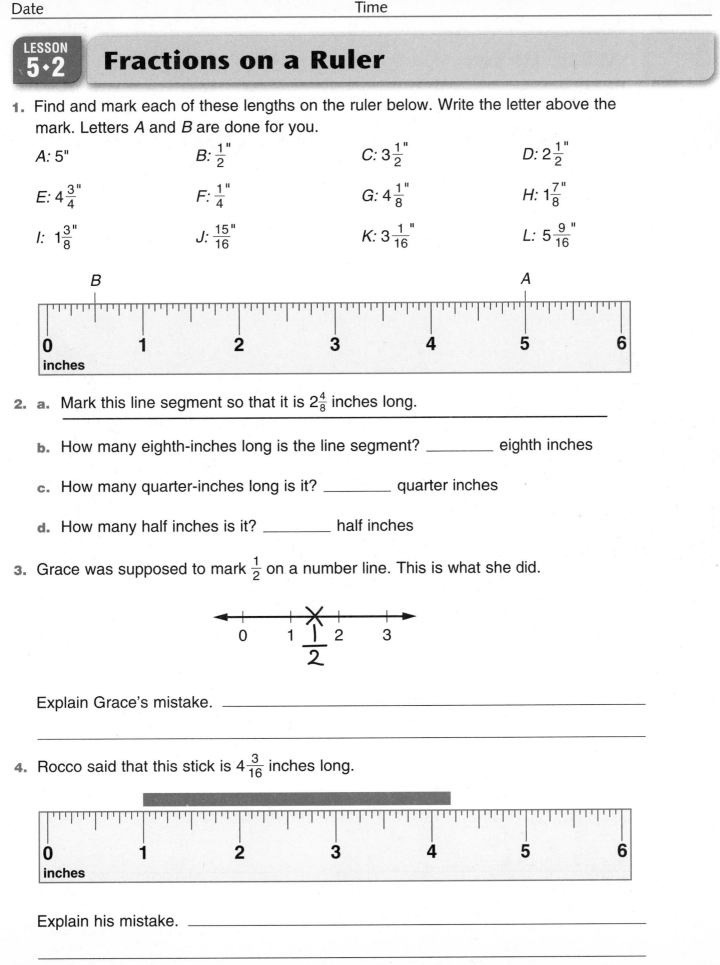

2. a. Mark this line segment so that it is $2\frac{4}{8}$ inches long.

   _____

   b. How many eighth-inches long is the line segment? _____ eighth inches

   c. How many quarter-inches long is it? _____ quarter inches

   d. How many half inches is it? _____ half inches

3. Grace was supposed to mark $\frac{1}{2}$ on a number line. This is what she did.

   Explain Grace's mistake. _____

   _____

4. Rocco said that this stick is $4\frac{3}{16}$ inches long.

   Explain his mistake. _____

Math Boxes

1. Write five fractions that are equivalent to $\frac{1}{2}$.

_____ _____ _____ _____ _____

SRB
77 78

2. Use a full circle protractor to draw and label an angle *MAD,* whose measure is 105°.

SRB
204–206

3. Raphael bought 14 pounds of meat to make hamburgers at the Fourth of July barbeque. He can make 5 hamburgers from each pound. Buns come in packages of 8. How many packages of buns does Raphael need?

_____
(unit)

Explain your answer.

_____

_____

_____

SRB
19 20
243

4. True or false? Write T or F.

a. 5,894 is divisible by 6. _____

b. 6,789 is divisible by 2. _____

c. 367 is divisible by 3. _____

d. 9,024 is divisible by 4. _____

e. 8,379 is divisible by 9. _____

SRB
11

5. Ella reads about 48 pages in 1 hour. About how many pages will she read in $2\frac{1}{2}$ hours? _____

SRB
59–61

LESSON 5·3   Comparing and Ordering Fractions

Math Message

Decide whether each of these measurements is closer to 0, $\frac{1}{2}$, or 1 inch. Circle the closest measurement.

1. $\frac{1}{8}$ inch is closest to . . .   0 inches. $\frac{1}{2}$ inch.   1 inch.

2. $\frac{15}{16}$ inch is closest to . . .   0 inches. $\frac{1}{2}$ inch.   1 inch.

3. $\frac{5}{8}$ inch is closest to . . .   0 inches. $\frac{1}{2}$ inch.   1 inch.

4. $\frac{3}{8}$ inch is closest to . . .   0 inches. $\frac{1}{2}$ inch.   1 inch.

5. Explain your solution for Problem 4.

_____

_____

Ordering Fractions

For each problem below, write the fractions in order from least to greatest.

6. $\frac{6}{8}, \frac{3}{8}, \frac{5}{8}, \frac{8}{8}$ _____, _____, _____, _____

7. $\frac{2}{7}, \frac{2}{9}, \frac{2}{5}, \frac{2}{12}$ _____, _____, _____, _____

8. $\frac{2}{3}, \frac{1}{4}, \frac{1}{3}, \frac{3}{4}$ _____, _____, _____, _____

9. $\frac{3}{5}, \frac{4}{10}, \frac{9}{20}, \frac{1}{25}$ _____, _____, _____, _____

10. $\frac{3}{7}, \frac{1}{10}, \frac{7}{8}, \frac{5}{7}$ _____, _____, _____, _____

11. $\frac{5}{9}, \frac{2}{5}, \frac{1}{6}, \frac{9}{10}$ _____, _____, _____, _____

12. $\frac{4}{8}, \frac{4}{7}, \frac{3}{5}, \frac{4}{9}$ _____, _____, _____, _____

LESSON 5·3   Fraction-Stick Chart

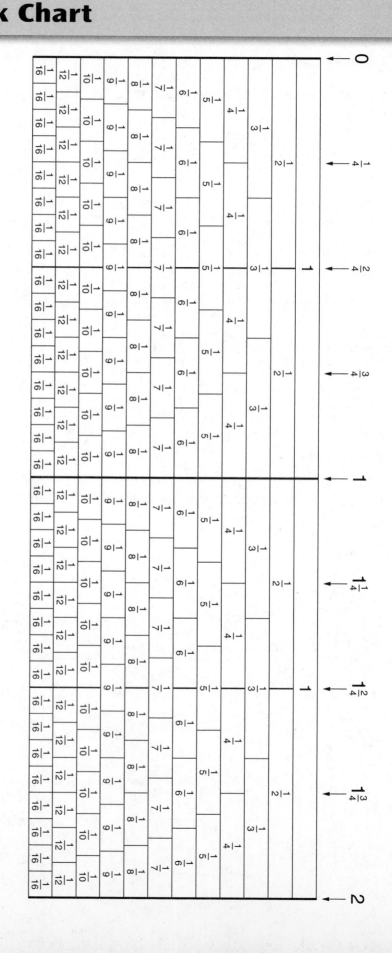

Fill in the blanks. Circle the correct answer.

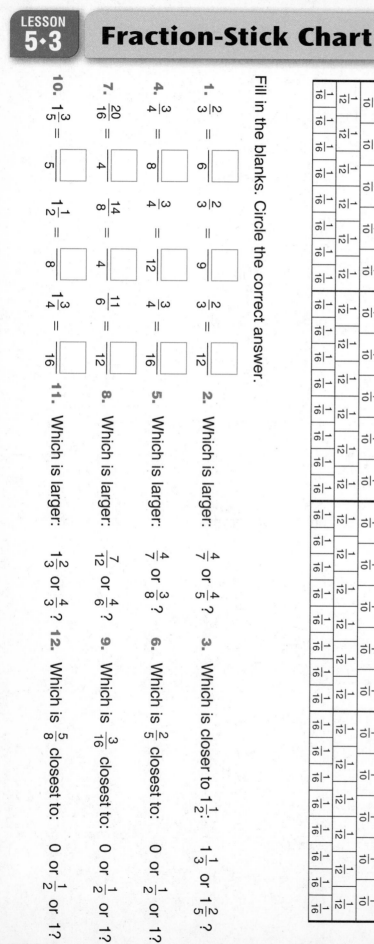

1. $\frac{2}{3} = \frac{\square}{6}$ $\frac{2}{3} = \frac{\square}{9}$ $\frac{2}{3} = \frac{\square}{12}$

2. Which is larger: $\frac{4}{7}$ or $\frac{4}{5}$?

3. Which is closer to $1\frac{1}{2}$: $1\frac{1}{3}$ or $1\frac{2}{5}$?

4. $\frac{3}{4} = \frac{\square}{8}$ $\frac{3}{4} = \frac{\square}{12}$ $\frac{3}{4} = \frac{\square}{16}$

5. Which is larger: $\frac{4}{7}$ or $\frac{3}{8}$?

6. Which is $\frac{2}{5}$ closest to: 0 or $\frac{1}{2}$ or 1?

7. $\frac{20}{16} = \frac{\square}{4}$ $\frac{14}{8} = \frac{\square}{4}$ $\frac{11}{6} = \frac{\square}{12}$

8. Which is larger: $\frac{7}{12}$ or $\frac{4}{6}$?

9. Which is $\frac{3}{16}$ closest to: 0 or $\frac{1}{2}$ or 1?

10. $1\frac{3}{5} = \frac{\square}{5}$ $1\frac{1}{2} = \frac{\square}{8}$ $1\frac{3}{4} = \frac{\square}{16}$

11. Which is larger: $1\frac{2}{3}$ or $\frac{4}{3}$?

12. Which is $\frac{5}{8}$ closest to: 0 or $\frac{1}{2}$ or 1?

LESSON 5·3

Adding with Fraction Sticks

A whole stick is worth 1.   [ ] = 1

[ ] = 2 halves   [ ] = 8 eighths

[ ] = 4 quarters   [ ] = 16 sixteenths

1. Use the fraction sticks to find equivalent fractions.

a. $\frac{1}{8} = \frac{\square}{16}$

b. $\frac{\square}{8} = \frac{12}{16} = \frac{\square}{4}$

c. $\frac{\square}{8} = \frac{3}{4} = \frac{\square}{16}$

d. $\frac{1}{2} = \frac{\square}{4} = \frac{\square}{8} = \frac{\square}{16}$

e. $\frac{\square}{2} = \frac{4}{4} = \frac{\square}{8} = \frac{\square}{16}$

2. Use the fraction sticks to add fractions with the same denominator.

Example: $\frac{1}{8} + \frac{2}{8} =$ [ ] $= \frac{3}{8}$

a. $\frac{2}{4} + \frac{1}{4} =$ [ ] $=$ _____

b. $\frac{3}{16} + \frac{9}{16} =$ [ ] $=$ _____

c. $\frac{1}{16} + \frac{5}{16} + \frac{8}{16} =$ [ ] $=$ _____

3. Use the fraction sticks to add fractions with different denominators.

a. $\frac{1}{2} + \frac{1}{4} =$ [ ] $=$ _____

b. $\frac{1}{2} + \frac{3}{8} =$ [ ] $=$ _____

c. $\frac{5}{8} + \frac{1}{4} =$ [ ] $=$ _____

d. $\frac{1}{4} + \frac{7}{8} + \frac{2}{16} =$ [ ] [ ] $=$ _____

LESSON 5·3 Fraction Number Stories

Shade the fraction sticks to help you solve these fraction number stories.
Write a number model for each story.

1. Chris made pizza dough with $\frac{5}{8}$ cup of white flour and $\frac{1}{4}$ cup of whole wheat flour.

    a. How much flour did he use in all? _____ cup

    b. Number model: _____

2. Sheryl's puppy weighed $1\frac{1}{2}$ pounds
   when it was born. After two weeks,
   the puppy had gained $\frac{3}{8}$ pounds.

    a. How much did the puppy weigh after two weeks? _____ pounds

    b. Number model: _____

3. Shade the fraction sticks to solve the number model. Then write a fraction number
   story that fits the number model.

    a. $\frac{3}{4} + \frac{5}{8} =$ _____

    b. Number story: _____

    _____

4. Make up your own fraction number story. Draw and shade fraction sticks to solve it.
   Write a number model for your story.

    a. Number story: _____

    _____

    b. Number model: _____

    c. Solution: _____

LESSON 5·3 Math Boxes

1. Write a 10-digit numeral that has

9 in the tens place,
3 in the millions place,
5 in the billions place,
7 in the hundred-millions place,
1 in the thousands place, and
6 in all other places.

___, ___ ___ ___, ___ ___ ___, ___ ___ ___

Write the numeral in words.

_____

_____

_____

_____

SRB 4

2. Write each fraction as a whole number or a mixed number.

a. $\frac{17}{4} =$ _____

b. $\frac{24}{3} =$ _____

c. $\frac{5}{2} =$ _____

d. $\frac{9}{8} =$ _____

e. $\frac{32}{5} =$ _____

SRB 62 63

3. Where 32 students vacationed, ...

a. what fraction of the students traveled within their state? _____

b. what fraction traveled to Europe? _____

c. what fraction traveled to Canada or Mexico? _____

Vacation Travel

within state 11
stayed home 3
Canada or Mexico 8
another state 6
Europe 4

SRB 59 125

4. Divide.

a. $21\overline{)493} \rightarrow$ _____

b. $35\overline{)623} \rightarrow$ _____

SRB 22–24

5. Find the following landmarks for this set of numbers: 929, 842, 986, 978, 869, 732, 898, 986, 900, 899, 986, 920, 842

a. Minimum: _____

b. Maximum: _____

c. Mode: _____

d. Range: _____

e. Median: _____

SRB 119

LESSON 5·4 Making Equivalent Fractions

Here is a way to model equivalent fractions. Start with a fraction stick that shows 3 out of 7 parts ($\frac{3}{7}$) shaded.

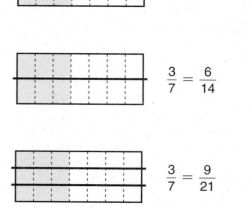

$\frac{3}{7}$

Draw a horizontal line to split each part of the stick into 2 equal parts. Now 6 out of 14 parts ($\frac{6}{14}$) are shaded. So $\frac{3}{7} = \frac{6}{14}$.

$\frac{3}{7} = \frac{6}{14}$

If each part of the original fraction stick is split into 3 equal parts, 9 out of 21 parts ($\frac{9}{21}$) are shaded. So $\frac{3}{7} = \frac{9}{21}$.

$\frac{3}{7} = \frac{9}{21}$

1. Draw horizontal lines to split each part of each fraction stick into 2 equal parts. Then fill in the missing numbers.

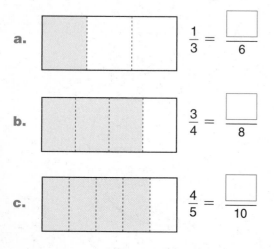

a. $\frac{1}{3} = \frac{\boxed{}}{6}$

b. $\frac{3}{4} = \frac{\boxed{}}{8}$

c. $\frac{4}{5} = \frac{\boxed{}}{10}$

2. Draw horizontal lines to split each part of each fraction stick into 3 equal parts. Then fill in the missing numbers.

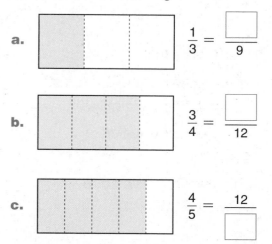

a. $\frac{1}{3} = \frac{\boxed{}}{9}$

b. $\frac{3}{4} = \frac{\boxed{}}{12}$

c. $\frac{4}{5} = \frac{12}{\boxed{}}$

3. Draw horizontal lines to split each part of each fraction stick into 4 equal parts. Then fill in the missing numbers.

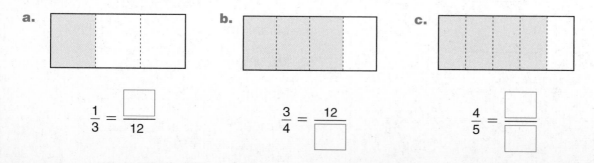

a. $\frac{1}{3} = \frac{\boxed{}}{12}$

b. $\frac{3}{4} = \frac{12}{\boxed{}}$

c. $\frac{4}{5} = \frac{\boxed{}}{\boxed{}}$

LESSON 5·4 # Finding Equivalent Fractions

Study the example below. Then solve Problems 1–3 in the same way. Match each fraction in the left column with an equivalent fraction in the right column.

Then fill in the boxes on the left with either a multiplication or division symbol and a number showing how you changed each fraction to get an equivalent fraction.

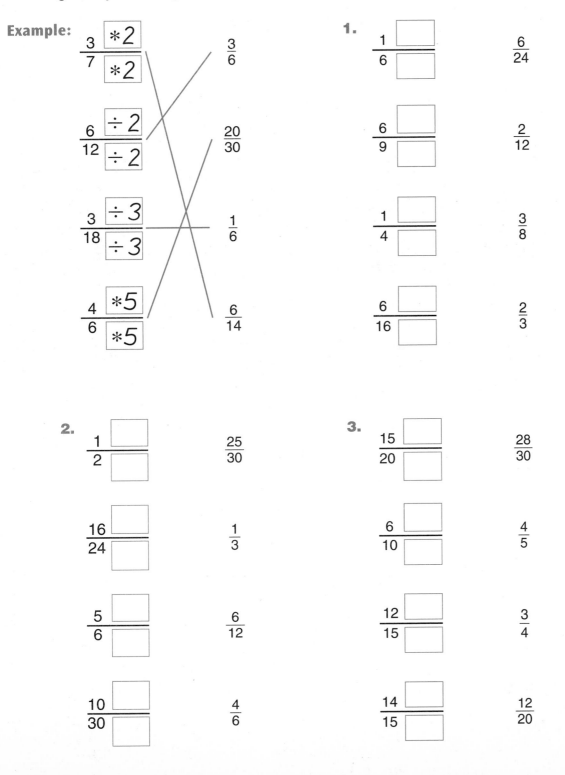

Example:

$\dfrac{3 \boxed{*2}}{7 \boxed{*2}}$ $\dfrac{3}{6}$

$\dfrac{6 \boxed{\div 2}}{12 \boxed{\div 2}}$ $\dfrac{20}{30}$

$\dfrac{3 \boxed{\div 3}}{18 \boxed{\div 3}}$ $\dfrac{1}{6}$

$\dfrac{4 \boxed{*5}}{6 \boxed{*5}}$ $\dfrac{6}{14}$

1.

$\dfrac{1 \boxed{}}{6 \boxed{}}$ $\dfrac{6}{24}$

$\dfrac{6 \boxed{}}{9 \boxed{}}$ $\dfrac{2}{12}$

$\dfrac{1 \boxed{}}{4 \boxed{}}$ $\dfrac{3}{8}$

$\dfrac{6 \boxed{}}{16 \boxed{}}$ $\dfrac{2}{3}$

2.

$\dfrac{1 \boxed{}}{2 \boxed{}}$ $\dfrac{25}{30}$

$\dfrac{16 \boxed{}}{24 \boxed{}}$ $\dfrac{1}{3}$

$\dfrac{5 \boxed{}}{6 \boxed{}}$ $\dfrac{6}{12}$

$\dfrac{10 \boxed{}}{30 \boxed{}}$ $\dfrac{4}{6}$

3.

$\dfrac{15 \boxed{}}{20 \boxed{}}$ $\dfrac{28}{30}$

$\dfrac{6 \boxed{}}{10 \boxed{}}$ $\dfrac{4}{5}$

$\dfrac{12 \boxed{}}{15 \boxed{}}$ $\dfrac{3}{4}$

$\dfrac{14 \boxed{}}{15 \boxed{}}$ $\dfrac{12}{20}$

LESSON 5·4   Math Boxes

1. Write five fractions that are equivalent to $\frac{3}{4}$.

_____ _____ _____ _____ _____

SRB
59

2. Use a full-circle protractor to draw and label the following angle:

∠*TOE:* 48°

SRB
204–206

3. Amanda found a can containing 237 dominoes. A full set has 28 dominoes. What is the greatest number of complete sets that can be found in the can? Explain how you found your answer.

_____

_____

_____

_____

SRB
22–24
243 246

4. True or False? Write T or F.

a. 1,704 is divisible by 4. _____

b. 7,152 is divisible by 6. _____

c. 8,264 is divisible by 3. _____

d. 4,005 is divisible by 2. _____

e. 2,793 is divisible by 9. _____

SRB
11

5. Donovan runs about 2.5 miles each day. In 3 weeks, about how many miles does he run?

Open sentence:

_____

Answer:

_____

SRB
38–40
219

LESSON 5·5   Math Message

Mark and label 3 decimals on each number line below.

1.

    0.7   0.8

2.

    9.32   9.33

Writing Fractions and Decimals

Write the numbers that your teacher dictates in the first column. Use the second column to show how you change these numbers to equivalent decimals. Write the decimal in the third column.

Fraction or Mixed Number	Calculations	Equivalent Decimals

LESSON 5·5 Renaming Fractions as Decimals

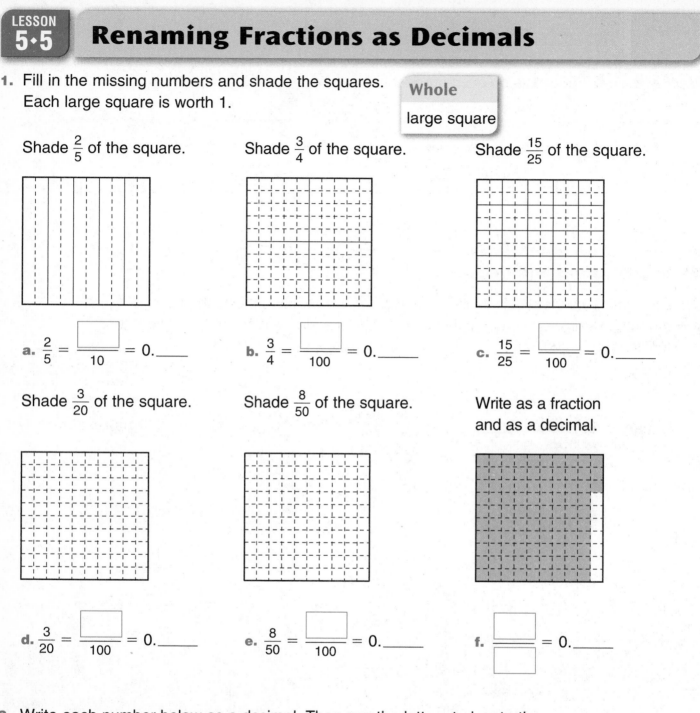

1. Fill in the missing numbers and shade the squares.
   Each large square is worth 1.

   Whole

   large square

   Shade $\frac{2}{5}$ of the square.

   Shade $\frac{3}{4}$ of the square.

   Shade $\frac{15}{25}$ of the square.

   a. $\frac{2}{5} = \frac{\boxed{}}{10} = 0.\underline{}$

   b. $\frac{3}{4} = \frac{\boxed{}}{100} = 0.\underline{}$

   c. $\frac{15}{25} = \frac{\boxed{}}{100} = 0.\underline{}$

   Shade $\frac{3}{20}$ of the square.

   Shade $\frac{8}{50}$ of the square.

   Write as a fraction and as a decimal.

   d. $\frac{3}{20} = \frac{\boxed{}}{100} = 0.\underline{}$

   e. $\frac{8}{50} = \frac{\boxed{}}{100} = 0.\underline{}$

   f. $\frac{\boxed{}}{\boxed{}} = 0.\underline{}$

2. Write each number below as a decimal. Then use the letters to locate the decimals on the number line.

   a. $\frac{1}{2} = \underline{0}.\underline{5}$

   b. $\frac{6}{10} = \underline{}.\underline{}$

   c. $\frac{4}{5} = \underline{}.\underline{}$

   d. $\frac{23}{100} = \underline{}.\underline{}$

   e. $\frac{22}{25} = \underline{}.\underline{}$

   f. $\frac{21}{50} = \underline{}.\underline{}$

   g. $\frac{7}{5} = \underline{}.\underline{}$

   h. $1\frac{15}{50} = \underline{}.\underline{}$

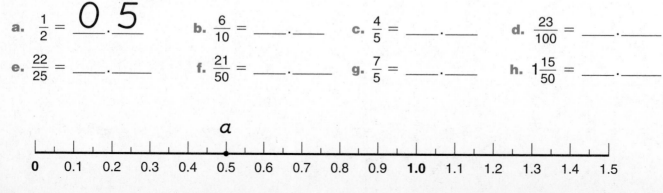

| 0 | 0.1 | 0.2 | 0.3 | 0.4 | 0.5 | 0.6 | 0.7 | 0.8 | 0.9 | 1.0 | 1.1 | 1.2 | 1.3 | 1.4 | 1.5 |

LESSON 5·5 Rounding Decimals

Rounding removes extra digits from a number. Sometimes decimals are shown with more digits than are useful. Data with too many decimal places can be misleading. For example, measurements with too many digits might seem more precise than they actually are. Many calculators display results to eight or more decimal places. In most situations, one or two places are enough.

Example:

The interest earned on a savings account at a bank is calculated to the nearest tenth of a cent. But the bank can't pay a fraction of a cent. Suppose one bank always *rounds* the interest down, and ignores any fraction of a cent.

The First Community Bank calculates interest on one account as $17.218 (17 dollars and 21.8 cents). The bank ignores the 0.8 (or $\frac{8}{10}$) cent. It pays $17.21 interest.

1. Here is the calculated monthly interest on Mica's First Community Bank savings account. Round each amount down to find the interest paid each month.

   | January | $21.403 | _____ | February | $22.403 | _____ |

   | March | $18.259 | _____ | April | $19.024 | _____ |

   | May | $17.427 | _____ | June | $18.916 | _____ |

   How much total interest did the bank pay Mica for these 6 months?

   (Add the rounded amounts.) _____

Example:

At the Olympic Games, each running event is timed to the nearest thousandth of a second. The timer *rounds* the time *up* to the *next* hundredth of a second (not the *nearest* hundredth). The rounded time becomes the official time.

11.437 seconds is rounded up to 11.44 seconds.

11.431 seconds is rounded up to 11.44 seconds.

11.430 seconds is reported as 11.43 seconds because 11.430 is equal to 11.43.

Michael Johnson with his record-breaking time

LESSON 5·5 Rounding Decimals *continued*

2. Find the official times for these runs.   min: minute(s)   s: second(s)

Electric Timer	Official Time	Electric Timer	Official Time
10.752 s	____ ____. ____ ____ s	20.001 s	____ ____. ____ ____ s
11.191 s	____ ____. ____ ____ s	43.505 s	____ ____. ____ ____ s
10.815 s	____ ____. ____ ____ s	49.993 s	____ ____. ____ ____ s
21.970 s	____ ____. ____ ____ s	1 min 55.738 s	____ min ____ ____. ____ s
20.092 s	____ ____. ____ ____ s	1 min 59.991 s	____ min ____ ____ s

Example:

Supermarkets often show unit prices for items. This helps customers comparison shop. A unit price is found by dividing the price of an item (in cents, or dollars and cents) by the quantity of the item (often in ounces or pounds). When the quotient has more decimal places than are needed, it is *rounded to the nearest* tenth of a cent.

23.822 cents (per ounce) is rounded down to 23.8 cents.

24.769 cents is rounded up to 24.8 cents.

18.65 cents is halfway between 18.6 cents and 18.7 cents. It is rounded up to 18.7 cents.

3. Round these unit prices to the nearest tenth of a cent (per ounce).

a. 28.374¢ _____   b. 19.796¢ _____   c. 29.327¢ _____

d. 16.916¢ _____   e. 20.641¢ _____   f. 25.583¢ _____

g. 18.469¢ _____   h. 24.944¢ _____   i. 17.281¢ _____

j. 23.836¢ _____   k. 21.866¢ _____   l. 22.814¢ _____

4. Describe a situation involving money when the result of a computation might always be rounded up.

_____

_____

LESSON 5·5 Math Boxes

1. Write a 4-digit number that has

3 in the hundredths place,
5 in the tenths place,
6 in the thousandths place, and
2 in the ones place.

_____ . _____ _____ _____

Write this number in words.

_____

_____

SRB 28

2. Rename each fraction as a mixed number or a whole number.

a. $\frac{28}{4}$ = _____

b. $\frac{36}{6}$ = _____

c. $\frac{25}{12}$ = _____

d. $\frac{46}{8}$ = _____

e. $\frac{18}{5}$ = _____

SRB 62

3. Write each mixed number as an improper fraction.

a. $1\frac{3}{4}$ = _____

b. $3\frac{1}{2}$ = _____

c. $2\frac{7}{8}$ = _____

d. $4\frac{9}{5}$ = _____

e. $6\frac{1}{3}$ = _____

SRB 62

4. Write the following numbers in order from least to greatest.

5.03 $4\frac{7}{4}$ 5.3 $\frac{3}{15}$ $5\frac{2}{5}$

_____, _____, _____, _____, _____

SRB 32 66 89

5. One week a family ate $3\frac{1}{2}$ boxes of cereal. The next week they ate $1\frac{3}{4}$ boxes. How many boxes did they eat in the two weeks?

_____

SRB 70

6. Wilkin School 5th graders wanted to donate 5 boxes of canned food to their local food bank. They collected $3\frac{1}{5}$ boxes in 3 days. How many more boxes do they need to collect?

_____

SRB 71

LESSON 5·6 Writing Fractions as Decimals

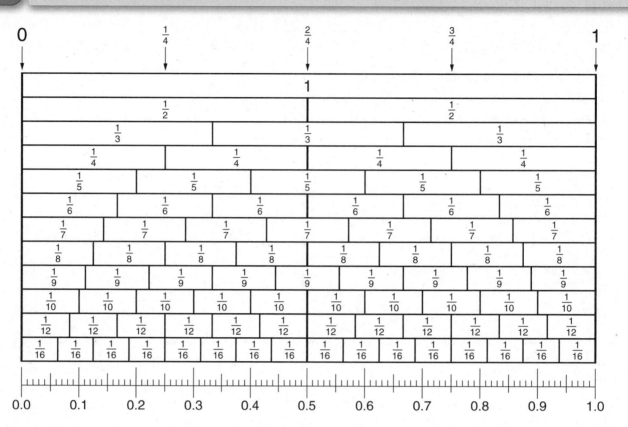

Use a straightedge and the above chart to fill in the blanks to the right of each fraction below. Write a decimal that is equal to, or about equal to, the given fraction. Directions for filling in the blank to the left of each fraction will be given in the next lesson.

1. _____ $\frac{1}{3}$ = 0.<u>3</u> <u>3</u>

2. _____ $\frac{2}{3}$ = 0.____ ____

3. _____ $\frac{4}{10}$ = 0.____ ____

4. _____ $\frac{4}{5}$ = 0.____ ____

5. _____ $\frac{1}{8}$ = 0.____ ____

6. _____ $\frac{5}{8}$ = 0.____ ____

7. _____ $\frac{9}{12}$ = 0.____ ____

8. _____ $\frac{11}{12}$ = 0.____ ____

9. _____ $1\frac{1}{3}$ = 1.<u>3</u> <u>3</u>

10. _____ $1\frac{3}{8}$ = 1.____ ____

11. _____ $3\frac{7}{8}$ = 3.____ ____

12. _____ $9\frac{5}{6}$ = ____.____ ____

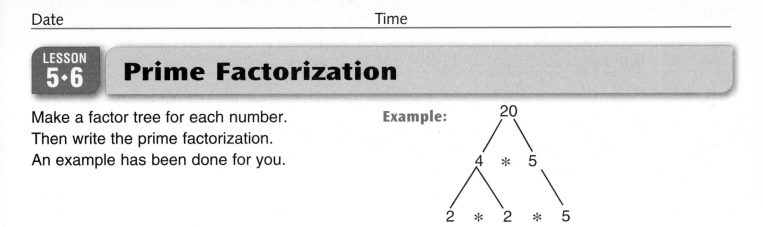

LESSON 5·6 Prime Factorization

Make a factor tree for each number.
Then write the prime factorization.
An example has been done for you.

Example:

$$20$$
$$4 * 5$$
$$2 * 2 * 5$$

1. 75

2. 99

3. 56

4. 32

Write each number as the product of factors. Then find the answer.

Example: $2^2 * 9 = 2 * 2 * 3 * 3 = 36$

5. $5^2 * 10^2 * 2^2 =$ _____ = _____

6. $3^2 * 2^3 * 10^2 =$ _____ = _____

7. $4^2 * 2^2 * 3^3 =$ _____ = _____

Write the prime factorization for each number using exponents when possible.

Example: $36 = 2^2 * 3^2$

8. $144 =$ _____   **9.** $1,000 =$ _____   **10.** $85 =$ _____

143

LESSON 5·6 Math Boxes

1. Complete the table.

Fraction	Decimal	Percent
$\frac{7}{10}$		
$\frac{3}{8}$		
	$0.\overline{3}$	

SRB
83 89
90

2. Estimate an answer for each problem.

a. $20.6 \div 4$ Estimate _____

b. $184.38 \div 9$ Estimate _____

c. $15.503 \div 7$ Estimate _____

d. $872.16 \div 8$ Estimate _____

SRB
42
247–249

3. Put the following decimals in order from least to greatest.

0.204

0.19

0.265

0.560

0.099

_____, _____, _____,

_____, _____

SRB
32 33

4. Determine whether the following sentences are true or false. Write T or F.

a. $3.5 > 0.35$ _____

b. $\frac{2}{5} < \frac{1}{3}$ _____

c. $2^4 = 64$ _____

d. $\sqrt{99} = 9$ _____

SRB
9 32 66

5. Put the following fractions in order from least to greatest.

$\frac{3}{8}, \frac{4}{5}, \frac{2}{3}, \frac{1}{4}, \frac{9}{10}$

____, ____, ____, ____, ____

SRB
66 67

6. Circle the letters for the pairs of equivalent fractions.

a. $\frac{1}{6}, \frac{3}{6}$ b. $\frac{15}{25}, \frac{3}{5}$

c. $\frac{2}{3}, \frac{6}{10}$ d. $\frac{2}{7}, \frac{10}{35}$

e. $\frac{48}{56}, \frac{6}{7}$ f. $\frac{4}{9}, \frac{65}{135}$

SRB
59–61

LESSON 5·7 More about Writing Fractions as Decimals

Sometimes when you divide without using the calculator's fix function to limit decimal places, you see that the display is filled with decimal digits. If the display were big enough, these digits would repeat forever.

Writing a bar over the digit or digits that repeat is a simple way to write these repeating decimals. Some calculators will display repeating decimals by rounding the last digit in the display. Study this table.

Fraction	Calculator Display	Decimal
$\frac{1}{3}$	0.33333333333	$0.\overline{3}$
$\frac{2}{3}$	0.66666666666 or 0.666666666667	$0.\overline{6}$
$\frac{1}{12}$	0.08333333333	$0.08\overline{3}$
$\frac{8}{9}$	0.8888888888 or 0.8888888888889	$0.\overline{8}$
$\frac{1}{22}$	0.045454545 or 0.045454545455	$0.0\overline{45}$

1. Explain how you would predict whether $\frac{2}{9}$ or $\frac{3}{9}$ is closer to 0.25 before using your calculator.

_____

_____

Use your calculator and convert each fraction below to a decimal by dividing the numerator by the denominator. If the result is a repeating decimal, write a bar over the digit or digits that repeat. Then circle the correct answer to each question.

2. Which is closer to 0.25? $\frac{2}{9}$ _____ or $\frac{3}{9}$ _____

3. Which is closer to 0.8? $\frac{6}{8}$ _____ or $\frac{5}{6}$ _____

4. Which is closer to 0.6? $\frac{4}{7}$ _____ or $\frac{7}{12}$ _____

5. Which is closer to 0.05? $\frac{1}{30}$ _____ or $\frac{1}{12}$ _____

6. Which is closer to 0.39? $\frac{3}{8}$ _____ or $\frac{7}{16}$ _____

LESSON 5·7 **Math Boxes**

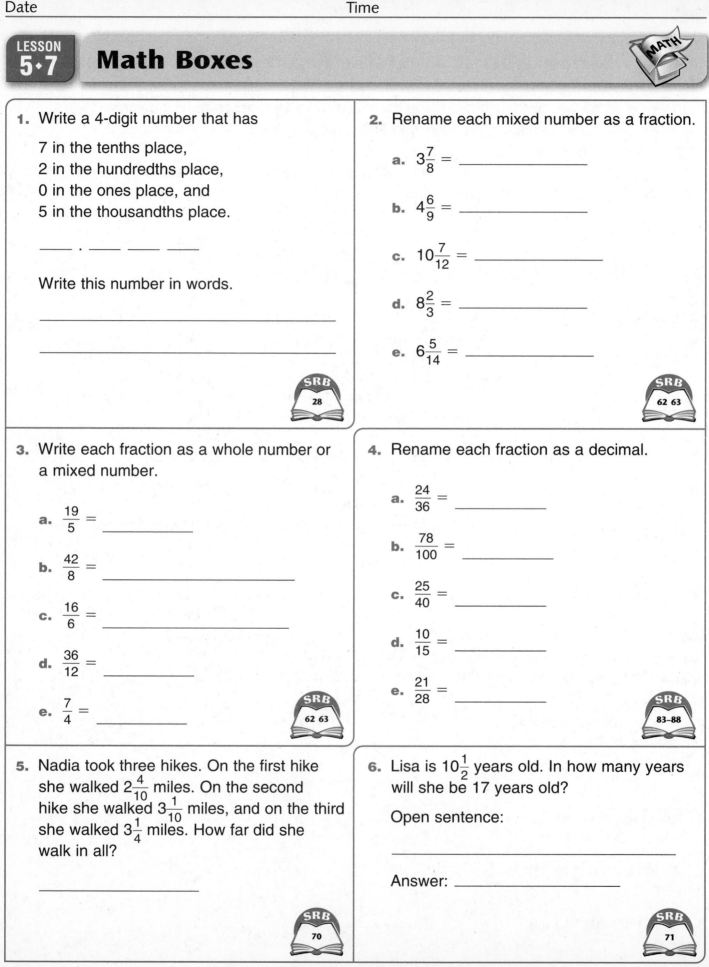

1. Write a 4-digit number that has

   7 in the tenths place,
   2 in the hundredths place,
   0 in the ones place, and
   5 in the thousandths place.

   ____ . ____ ____ ____

   Write this number in words.

   _____

   _____

   SRB
   28

2. Rename each mixed number as a fraction.

   a. $3\frac{7}{8}$ = _____

   b. $4\frac{6}{9}$ = _____

   c. $10\frac{7}{12}$ = _____

   d. $8\frac{2}{3}$ = _____

   e. $6\frac{5}{14}$ = _____

   SRB
   62 63

3. Write each fraction as a whole number or a mixed number.

   a. $\frac{19}{5}$ = _____

   b. $\frac{42}{8}$ = _____

   c. $\frac{16}{6}$ = _____

   d. $\frac{36}{12}$ = _____

   e. $\frac{7}{4}$ = _____

   SRB
   62 63

4. Rename each fraction as a decimal.

   a. $\frac{24}{36}$ = _____

   b. $\frac{78}{100}$ = _____

   c. $\frac{25}{40}$ = _____

   d. $\frac{10}{15}$ = _____

   e. $\frac{21}{28}$ = _____

   SRB
   83–88

5. Nadia took three hikes. On the first hike she walked $2\frac{4}{10}$ miles. On the second hike she walked $3\frac{1}{10}$ miles, and on the third she walked $3\frac{1}{4}$ miles. How far did she walk in all?

   _____

   SRB
   70

6. Lisa is $10\frac{1}{2}$ years old. In how many years will she be 17 years old?

   Open sentence:

   _____

   Answer: _____

   SRB
   71

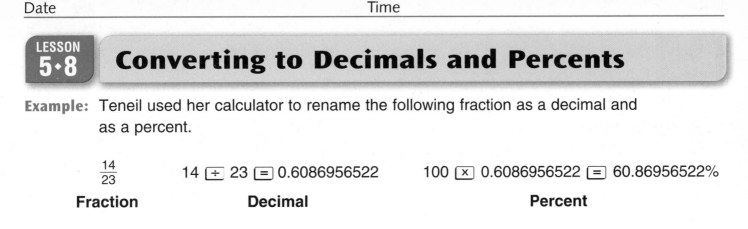

LESSON 5·8

Converting to Decimals and Percents

Example: Teneil used her calculator to rename the following fraction as a decimal and as a percent.

$\frac{14}{23}$ 14 ÷ 23 = 0.6086956522   100 × 0.6086956522 = 60.86956522%

Fraction   **Decimal**   **Percent**

Teneil needed to work with only a whole percent, so she rounded 60.86956522% to 61%.

1. Use your calculator to convert each fraction to a decimal. Write all of the digits shown in the display. Then write the equivalent percent rounded to the nearest whole percent. The first row has been done for you.

Fraction	Decimal	Percent
$\frac{18}{35}$	0.5142857143	51%
$\frac{12}{67}$		
$\frac{24}{93}$		
$\frac{13}{24}$		
$\frac{576}{1,339}$		

2. Lionel got 80% of the questions correct on a spelling test. If the test had 20 questions, how many did Lionel get correct? _____

3. Jamie spent 50% of his money on a baseball cap. The cap cost $15. How much money did Jamie have at the beginning? _____

4. Hunter got 75% of the questions correct on a music test. If he got 15 questions correct, how many questions were on the test? _____

LESSON 5·8   Converting to Decimals and Percents *cont.*

5. The chart lists 6 animals and the average number of hours per day that each spends sleeping.

Write the fraction of a day that each animal sleeps. Then calculate the equivalent decimal and percent, rounded to the nearest whole percent. You may use your calculator. The first row has been done for you.

Animal	Average Hours of Sleep per Day	Fraction of Day Spent Sleeping	Decimal Equivalent	Percent of Day Spent Sleeping
Lion and Sloth	20	$\frac{20}{24}$	$0.8\overline{3}$	83%
Opossum	19			
Armadillo and Koala	18			
Southern Owl Monkey	17			

Source: *The Top 10 of Everything 2005*

6. The total number of horses in the world is about 60,800,000. China is the country with the greatest number of horses (about 8,900,000). What percent of the world's horses live in China? _____

7. In the United States, about 45% of the population has blood type O. About how many people out of every 100 have blood type O? _____

8. About 11 out of every 100 households in the United States has a parakeet. How would you express this as a percent? _____

9. It is thought that adults need an average of 8 hours of sleep per day.

a. What percent of the day should adults sleep? _____

b. If infants average 16 hours of sleep per day, what percent of the day do they sleep? _____

LESSON 5·8   Math Boxes

1. Complete the table.

Fraction	Decimal	Percent
$\frac{2}{3}$		
	0.95	
		43%
$\frac{3}{5}$		
	0.8	

SRB 83 89 90

2. Estimate an answer for each problem.

a. $4\overline{)39.04}$ Estimate _____

b. $8\overline{)17.6}$ Estimate _____

c. $5\overline{)300.007}$ Estimate _____

SRB 42 247–250

3. Put the following decimals in order from least to greatest.

0.4, 0.14, 0.44, 0.415

_____, _____,

_____, _____

SRB 32 33

4. Determine whether the following sentences are true or false. Write T or F.

a. $6.05 + 2.7 = 8.75$ _____

b. If $n = 4$, then $3 * n = 12$ _____

c. $2^2 + 3^2 = 5^2$ _____

d. $\sqrt{64} = 8$ _____

e. $2^2 * \sqrt{100} = 40$ _____

SRB 9 32 33

5. Put the following fractions in order from least to greatest.

$\frac{3}{7}, \frac{3}{5}, \frac{2}{8}, \frac{8}{9}, \frac{5}{6}$

_____, _____, _____, _____, _____

SRB 66 67

6. Circle the letters for the pairs of equivalent fractions.

a. $\frac{4}{9}, \frac{16}{36}$

b. $\frac{18}{81}, \frac{2}{9}$

c. $\frac{28}{40}, \frac{14}{25}$

d. $\frac{10}{15}, \frac{5}{7}$

e. $\frac{5}{25}, \frac{1}{5}$

SRB 59–61

LESSON 5·9 Math Message

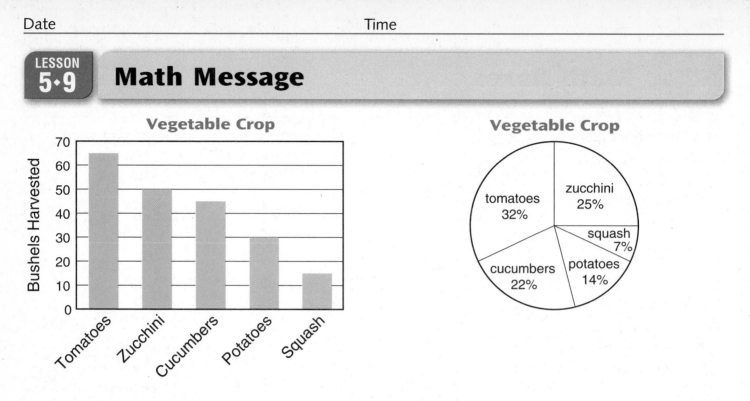

List how the 2 graphs above are the same and how they are different.

Similarities

Differences	
Bar Graph	**Circle Graph**

LESSON 5·9   Bar Graphs and Circle (Pie) Graphs

1. Circle the after-school snack you like best. Mark only one answer.

cookies   granola bar   candy bar   fruit   other

2. Record the class results of the survey.

cookies _____ granola bar _____ candy bar _____ fruit _____ other _____

Add all of the votes. Total: _____

The total is the number of students who voted.

3. Make and label a bar graph of the class data showing the results.

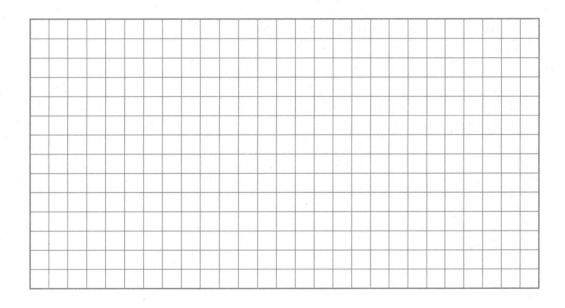

4. Another fifth-grade class with 20 students collected snack-survey data. The class made the circle graph (also called a pie graph) below.

Tell how you think they made the graph.

_____

_____

_____

Our Snacks

Math Boxes

1. Write a 4-digit number with

0 in the tenths place,
4 in the thousandths place,
1 in the hundredths place, and
5 in the ones place.

_____ . _____ _____ _____

Write this number in words.

_____

_____

SRB
28

2. Use the division rule to find equivalent fractions.

a. $\dfrac{\boxed{}}{8}$, or $\dfrac{1}{2} = \dfrac{2}{4}$ b. $\dfrac{9}{36} = \dfrac{\boxed{}}{12}$

c. $\dfrac{12}{144} = \dfrac{\boxed{}}{12}$ d. $\dfrac{14}{49} = \dfrac{2}{\boxed{}}$

e. $\dfrac{21}{63} = \dfrac{\boxed{}}{3}$, or $\dfrac{3}{9}$ f. $\dfrac{6}{14} = \dfrac{\boxed{}}{7}$

SRB
66 67

3. Use the information in the bar graph to complete the circle graph.

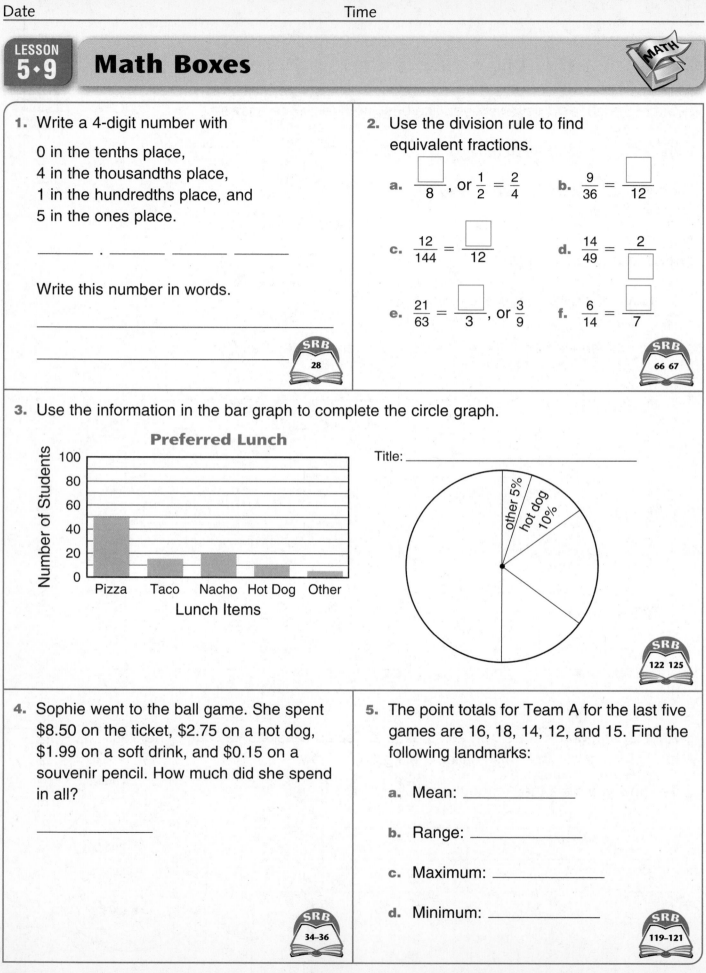

Preferred Lunch

Number of Students / Lunch Items
Pizza, Taco, Nacho, Hot Dog, Other

Title: _____

other 5%
hot dog 10%

SRB
122 125

4. Sophie went to the ball game. She spent $8.50 on the ticket, $2.75 on a hot dog, $1.99 on a soft drink, and $0.15 on a souvenir pencil. How much did she spend in all?

_____

SRB
34–36

5. The point totals for Team A for the last five games are 16, 18, 14, 12, and 15. Find the following landmarks:

a. Mean: _____

b. Range: _____

c. Maximum: _____

d. Minimum: _____

SRB
119–121

LESSON 5·10 Reading Circle Graphs

For each piece of the graph, estimate what fraction and what percent of the whole circle it represents. Label the graph pieces with the estimates. Then write your measured percents in the graph legend.

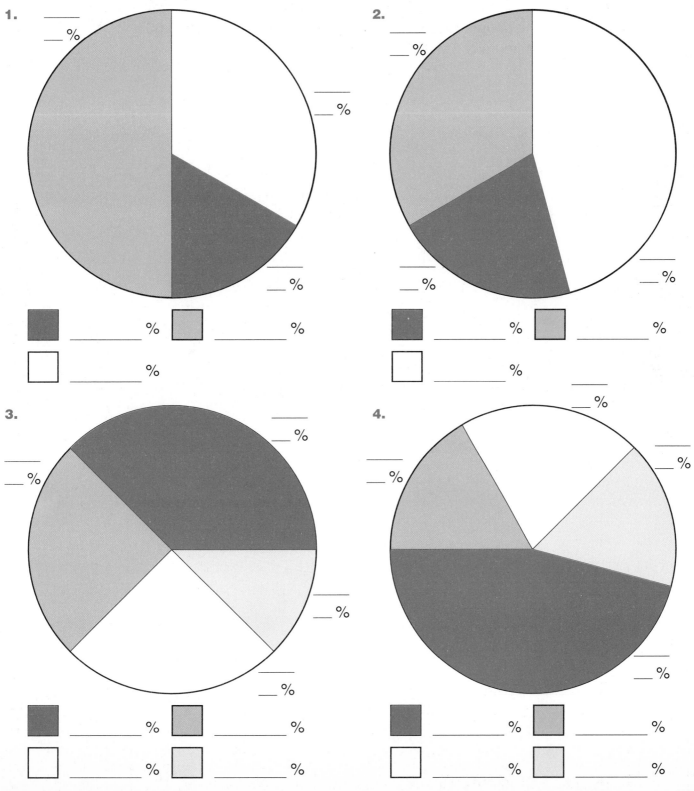

1. _____
       ___ %
                                ___ %

       _____
       ___ %

☐ _____ %   ☐ _____ %

☐ _____ %

2. _____
       ___ %
                                ___ %

       _____   _____
       ___ %   ___ %

☐ _____ %   ☐ _____ %

☐ _____ %

3. _____
       ___ %

       _____
       ___ %
                                ___ %

       _____
       ___ %

☐ _____ %   ☐ _____ %

☐ _____ %   ☐ _____ %

4.   ___ %

       _____   _____
       ___ %   ___ %

                                ___ %

☐ _____ %   ☐ _____ %

☐ _____ %   ☐ _____ %

LESSON 5·10 How Much TV Do People Watch?

A large sample of people was asked to report how much TV they watched during one week. The circle graph below shows the survey's results.

Estimate each percent. Then use your Percent Circle to measure the percent in each category. Write your measures in the blanks next to the pie pieces. Two percents are filled in for you: 18% of the people reported watching less than 7 hours, and 30% reported watching 7 to 14 hours of TV during the week.

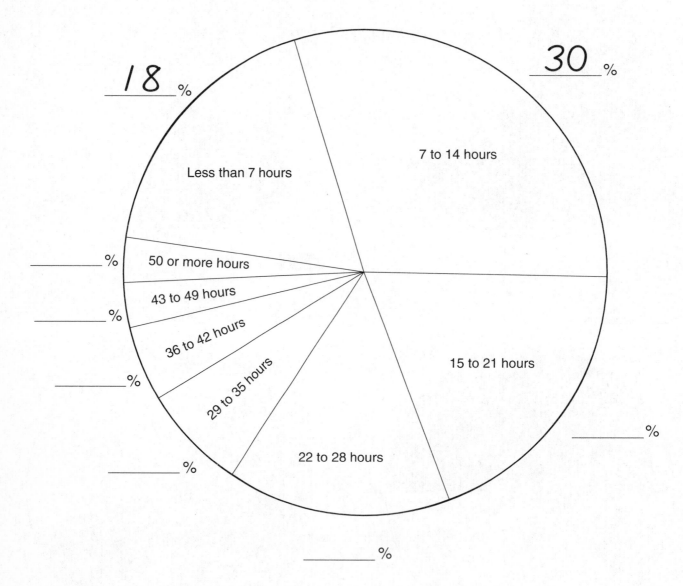

$\underline{} 18 $%

$\underline{} 30 $%

7 to 14 hours

Less than 7 hours

_____ %   50 or more hours

_____ %   43 to 49 hours

36 to 42 hours

_____ %

29 to 35 hours

15 to 21 hours

_____ %

_____ %

22 to 28 hours

_____ %

LESSON 5·10 **Measuring Angles, Perimeter, and Area**

1. Estimate the angle measure. Then use a protractor to measure the angle.

a.

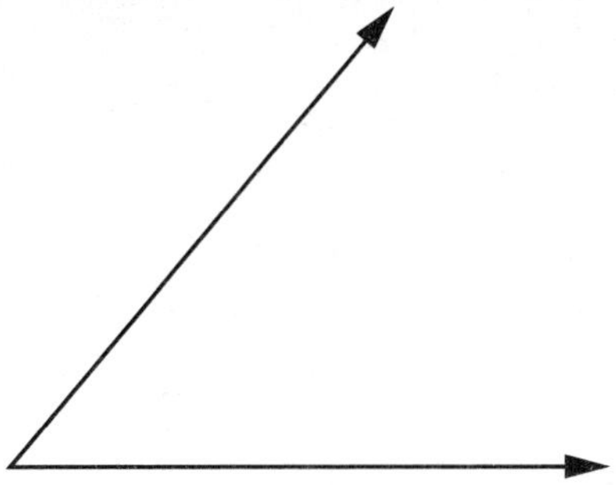

b.

Estimate: _____

Measure: _____

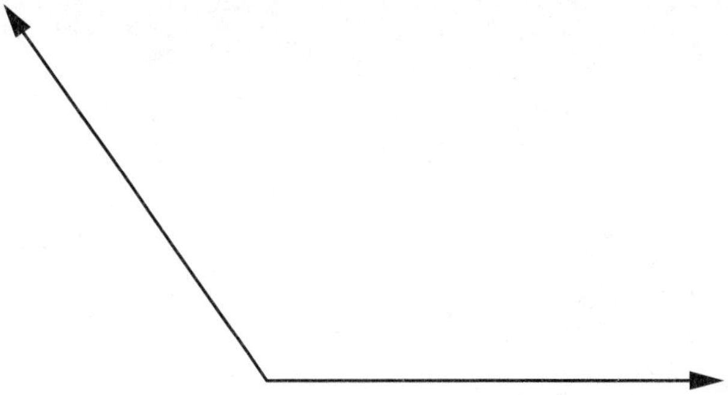

Estimate: _____

Measure: _____

2. Use a protractor to draw an angle with the given measure.

a. 75°

b. 145°

3. Find the missing measurement in each figure.

a.
14 in.
S

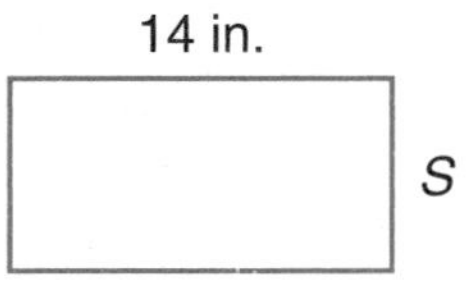

Perimeter = 42 in.

S = _____

b.
R

Perimeter = 28 m

R = _____

c.
17 m 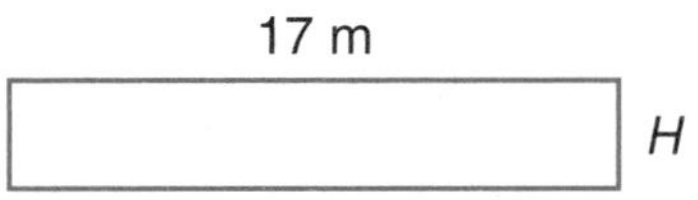
H

Area = 51 m²

H = _____

d.
13 cm 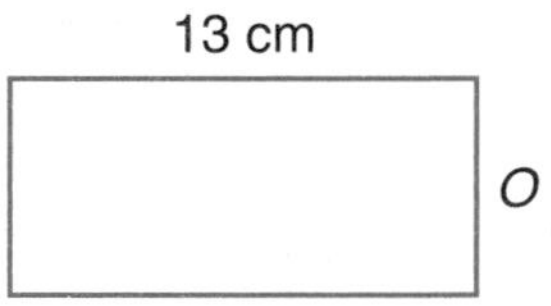
O

Area = 78 cm²

O = _____

4. The art room at Walker School is 27 feet long and 38 feet wide.

What is the area of the room? _____

5. Mia's backyard is 25 feet by 50 feet. Joy's backyard is 35 feet by 35 feet.

Who has a bigger backyard? _____ How much bigger? _____

LESSON 5·10 Math Boxes

1. Estimate.

a. $6.32 - 3.91 \rightarrow$ _____

b. $173.6 + 65.8 \rightarrow$ _____

c. $3.6 * 19 \rightarrow$ _____

d. $9.9 * 54 \rightarrow$ _____

e. $21.6 * 100 \rightarrow$ _____

SRB 247–250

2. Measure angle *SUM* to the nearest degree.

_____

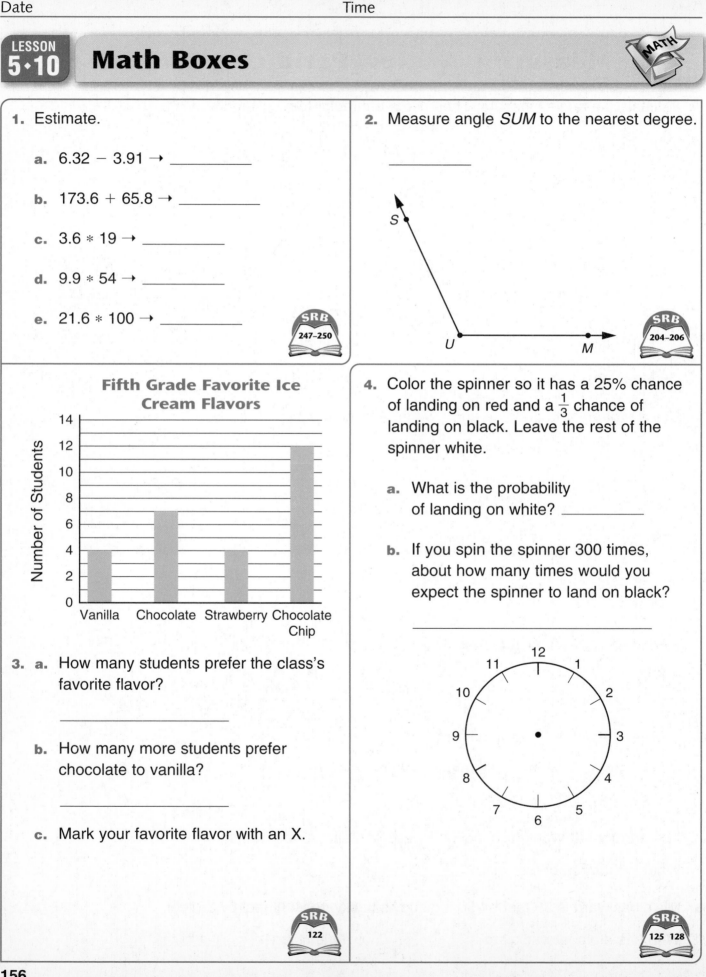

SRB 204–206

Fifth Grade Favorite Ice Cream Flavors

3. a. How many students prefer the class's favorite flavor?

_____

b. How many more students prefer chocolate to vanilla?

_____

c. Mark your favorite flavor with an X.

SRB 122

4. Color the spinner so it has a 25% chance of landing on red and a $\frac{1}{3}$ chance of landing on black. Leave the rest of the spinner white.

a. What is the probability of landing on white? _____

b. If you spin the spinner 300 times, about how many times would you expect the spinner to land on black?

_____

SRB 125 128

156

LESSON 5·11

Making Circle Graphs: Concrete Recipe

Concrete is an artificial stone. It is made by first mixing cement and sand with gravel or other broken stone. Then enough water is mixed in to cause the cement to set. After drying (or curing), the result is a hard piece of concrete.

The cement, sand, and gravel are commonly mixed using this recipe.

Recipe for Dry Mix for Concrete		
Material	**Fractional Part of Mix**	**Percent Part of Mix**
Cement	$\frac{1}{6}$	$16\frac{2}{3}\%$
Sand	$\frac{1}{3}$	$33\frac{1}{3}\%$
Gravel	$\frac{1}{2}$	50%

Use your Percent Circle to make a circle graph for the above recipe in the circle below. Label each sector of the graph and give it a title.

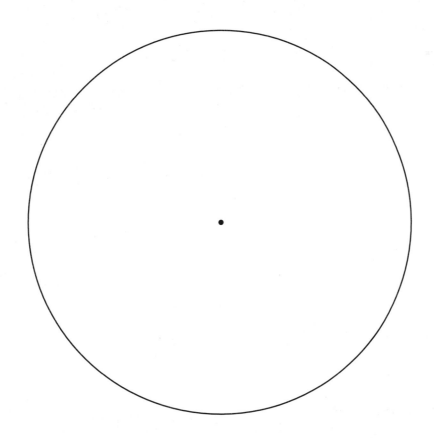

LESSON 5·11 Making Circle Graphs: Snack Survey

Your class recently made a survey of favorite snacks. As your teacher tells you the percent of votes each snack received, record the data in the table at the right. Make a circle graph of the snack-survey data in the circle below. Remember to label each piece of the graph and give it a title.

Votes			
Snack	Number	Fraction	Percent
Cookies			
Granola Bar			
Candy Bar			
Fruit			
Other			
Total			About 100%

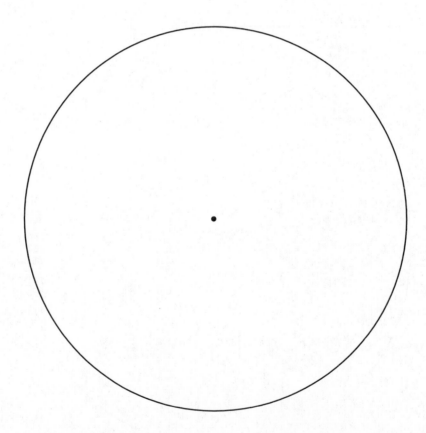

LESSON 5·11 Math Boxes

1. Write a 5-digit number with

   5 in the tens place,
   5 in the thousandths place, and
   0 in all the other places.

   _____ _____ . _____ _____ _____

   Write this number in words.

   _____

   _____

   SRB
   28

2. Use the division rule to find equivalent fractions in simplest form.

   a. $\frac{25}{100} =$ _____

   b. $\frac{8}{24} =$ _____

   c. $\frac{27}{36} =$ _____

   d. $\frac{16}{24} =$ _____

   SRB
   66 67

3. Use your Percent Circle and the information in the bar graph to complete the circle graph.

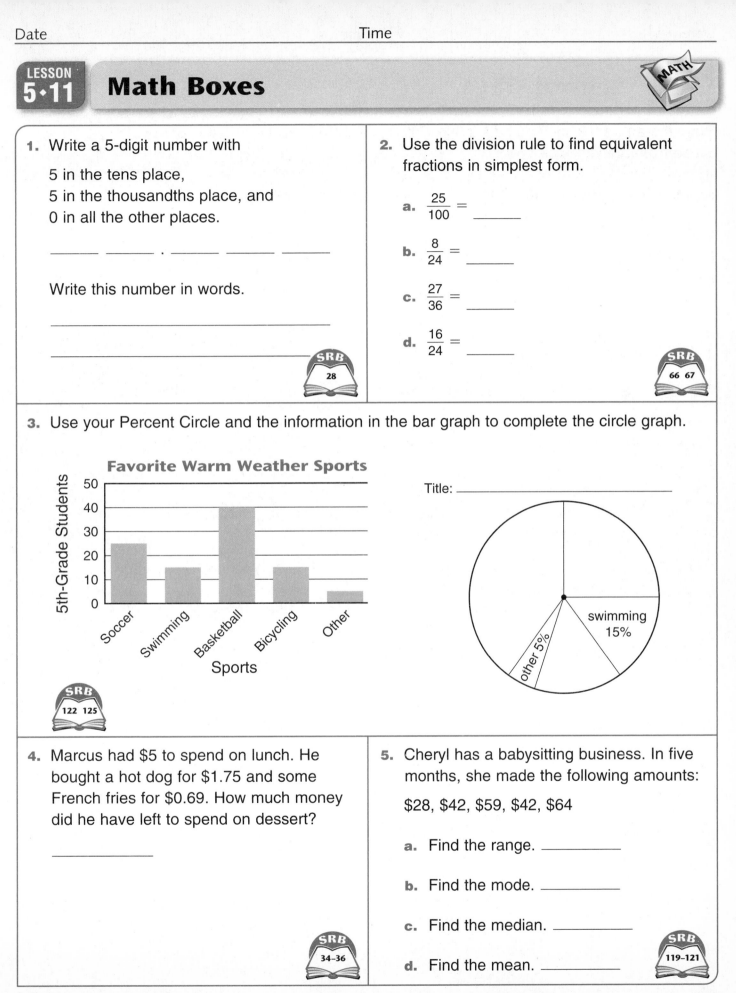

   Favorite Warm Weather Sports

   Title: _____

4. Marcus had $5 to spend on lunch. He bought a hot dog for $1.75 and some French fries for $0.69. How much money did he have left to spend on dessert?

   _____

   SRB
   34–36

5. Cheryl has a babysitting business. In five months, she made the following amounts:

   $28, $42, $59, $42, $64

   a. Find the range. _____

   b. Find the mode. _____

   c. Find the median. _____

   d. Find the mean. _____

   SRB
   119–121

159

LESSON 5·12 School Days

Read the article "School" on pages 360–362 in the American Tour section of the *Student Reference Book*.

1. Tell whether the statement below is true or false. Support your answer with evidence from page 360 of the American Tour.

   In 1790, it was common for 11-year-olds to go to school fewer than 90 days per year.

   _____

   _____

   _____

2. About how many days will you go to school this year? About _____ days

   Write a fraction to compare the number of days you will go to school this year to the number of days an 11-year-old might have gone to school in 1790.

   _____

3. Tell whether the statement below is true or false. Support your answer with evidence from page 361 of the American Tour.

   In 1900, students in some states spent twice as many days in school, on average, as students in some other states.

   _____

   _____

   _____

4. In 1900, in which region (Northeast, South, Midwest, or West) did students go to school …

   the greatest number of days per year? _____

   the fewest number of days per year? _____

LESSON 5·12   **School Days** *continued*

Tell whether each statement below is true or false. Support your answer with evidence from the graphs on page 362 of the American Tour.

5. On average, students in 2000 were absent from school about one-third as many days as students were absent in 1900.

_____

_____

_____

6. The average number of days students spent in school per year has not changed much since 1960.

_____

_____

_____

Try This

7. Tell whether the statement below is true or false. Support your answer with evidence from the American Tour.

From 1900 to 1980, the average number of days students spent in school per year more than doubled.

_____

_____

_____

8. Locate your state in the table Average Number of Days in School per Student, 1900 on page 361 of the American Tour. If you are in Alaska or Hawaii, choose another state.

Was your state above or below the median for its region? _____

9. Locate the number of days in school for your state in the stem-and-leaf plot on page 361 of the American Tour.

Was your state above or below the median for all states? _____

Math Boxes

1. Estimate. Show answers in whole numbers.

   a. $31.2 \div 9.6 \rightarrow$ _____

   b. $4.82 * 47 \rightarrow$ _____

   c. $159.5 + 27.9 \rightarrow$ _____

   d. $8.13 - 2.92 \rightarrow$ _____

   e. $62.1 * 3.2 \rightarrow$ _____

   SRB
   247–250

2. Measure each angle.

   a. $\angle LID$ measures

      about _____.

   b. $\angle FUN$ measures

      about _____.

   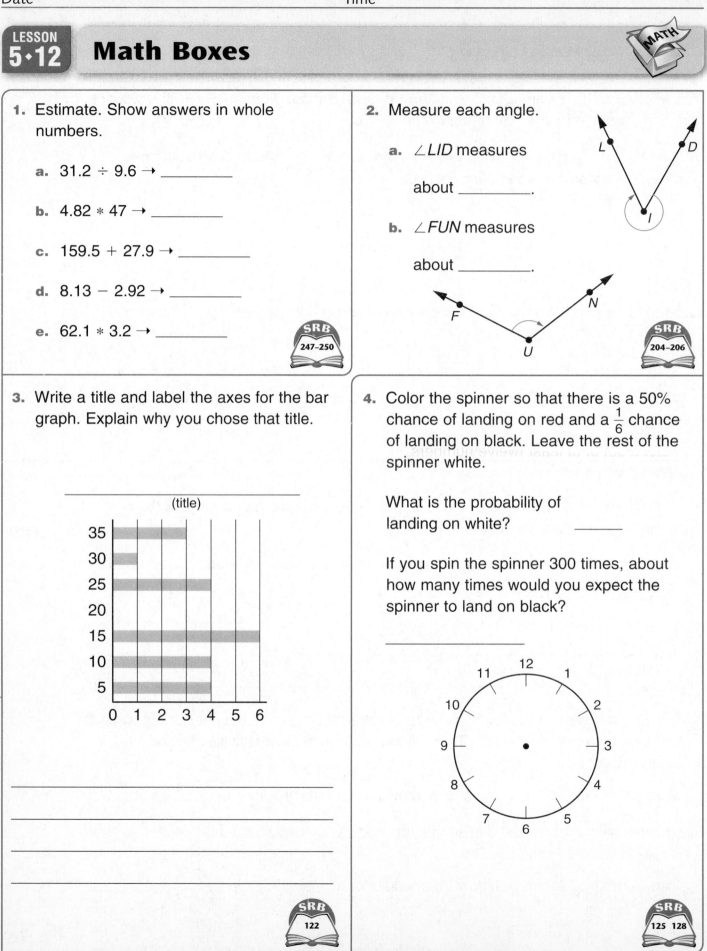

   SRB
   204–206

3. Write a title and label the axes for the bar graph. Explain why you chose that title.

   (title)

   35
   30
   25
   20
   15
   10
   5

   0 1 2 3 4 5 6

   _____

   _____

   _____

   _____

   SRB
   122

4. Color the spinner so that there is a 50% chance of landing on red and a $\frac{1}{6}$ chance of landing on black. Leave the rest of the spinner white.

   What is the probability of
   landing on white?   _____

   If you spin the spinner 300 times, about how many times would you expect the spinner to land on black?

   _____

   SRB
   125 128

LESSON 5·13 **Math Boxes**

1. Mr. Hernandez's class took a survey to find out when students prefer to do their homework. 125 fifth-grade students responded.

As soon as I get home	17%
After having an after-school snack	30%
Right after dinner	39%
Just before going to bed	14%

Use your Geometry Template to make a circle graph of the results. Give the graph a title. Label the sectors of the graph.

(title)

SRB
125 126

2. List a set of at least twelve numbers that has the following landmarks:

Minimum: 28   Maximum: 34
Median: 30   Mode: 29

_____

_____

Make a bar graph for what this set of numbers might represent.

34
33
32
31
30
29
28
0
Jan Feb Mar Apr May Jun Jul Aug Sep Oct Nov Dec

SRB
119 122

3. If you roll a regular 6-sided die, what is the probability of getting ...

a. 5? _____

b. a prime number? _____

c. an even number? _____

d. a multiple of 3? _____

SRB
128 129

4. Draw and label a 30° angle.

SRB
204–206

163

LESSON 6·1   States Students Have Visited

1. You and your classmates counted the number of states each of you has visited. As the counts are reported and your teacher records them, write them in the space below. When you finish, circle your own count in the list.

2. Decide with your group how to organize the data you just listed. (For example, you might make a line plot or a tally table.) Then organize the data and show the results below.

3. Write two things that you think are important about the data.

   a. _____

   b. _____

4. Compare your own count of states with those of your classmates.

   _____

   _____

LESSON 6·1   States Adults Have Visited

1. You and your classmates each recorded the number of states an adult has visited. As the numbers are reported and your teacher records them, write them in the space below.

2. Draw a line plot to organize the data you just listed.

3. How does the line plot help a viewer see what is important about the data?

_____

_____

_____

4. Record landmarks for the data about adults and students in the table below.

Landmark	Adults	Students
Minimum		
Maximum		
Mode(s)		
Median		

LESSON 6·1

Models for Rounding Numbers

Rounding numbers makes mental calculations and estimates easier. The first step to rounding to a given place is to locate the number between two consecutive numbers. For example, what whole number comes right before 12.6? (12) What whole number comes right after 12.6? (13)

1. Complete the table below.

	Rounding to the nearest whole number:	Rounding to the nearest ten:
12.6	Is between 12 and 13	Is between 10 and 20
26.3	Is between ___ and ___	Is between ___ and ___

	Rounding to the nearest ten:	Rounding to the nearest hundred:
119.9	Is between _____ and _____	Is between _____ and _____
3,502	Is between _____ and _____	Is between _____ and _____

Next determine if the number you are rounding is closer to the lower number or to the higher number. The models below represent different ways of thinking about making this choice. On the number line, 12.6 is close to 13, so 12.6 rounded to the nearest whole number is 13. If the number line were curved, 12.6 would "roll" toward 10.

Rounding to the nearest whole number

Rounding to the nearest ten

Numbers that are halfway between the lower number and the higher number are rounded up to the higher number. Round . . .

2. 16.4 to the nearest whole number. _____

3. 482 to the nearest hundred. _____

4. 7.36 to the nearest tenth. _____

5. 9,282 to the nearest hundred. _____

6. 423,897 to the nearest hundred-thousand. _____

7. 30.08 to the nearest tenth. _____

8. Explain how you would use your calculator to round 5.3458 to the nearest hundredth.

_____

_____

LESSON 6·1   Math Boxes

1. Fill in the missing values on the number lines.

29 _____ _____ _____ 57 _____ 71

19 _____ 53 _____ _____ _____ _____

SRB 230–233

2. Make up a set of at least twelve numbers that have the following landmarks.

minimum: 2
maximum: 11
median: 6
mode: 6

_____

_____

Make a bar graph for this set of numbers.

SRB 119 122

3. Write 3 equivalent fractions for each fraction.

a. $\dfrac{80}{100} =$ _____

b. $\dfrac{2}{3} =$ _____

c. $\dfrac{36}{9} =$ _____

d. $\dfrac{3}{24} =$ _____

e. $\dfrac{3}{8} =$ _____

SRB 59

4. Write each number in standard notation.

a. $3^3 =$ _____

b. $10^3 =$ _____

c. $6^3 =$ _____

d. $4 * 10^3 =$ _____

e. $7^2 =$ _____

SRB 6 8

LESSON 6·2

Personal Measures

Reference
10 millimeters (mm) = 1 centimeter (cm)
100 centimeters = 1 meter (m)
1,000 millimeters = 1 meter
1 inch (in.) is equal to about $2\frac{1}{2}$ (2.5) centimeters.

Work with a partner. You will need a ruler and a tape measure. Both tools should have metric units (millimeters and centimeters) and U.S. customary units (inches).

Find your own personal measures for each body unit shown. First measure and record using metric units. Then measure and record using U.S. customary units. Measure the lengths in Problems 1–4 to the nearest millimeter and $\frac{1}{16}$ in. and Problems 6–8 to the nearest centimeter and $\frac{1}{4}$ in.

1. **1-finger width**

   _____ mm

   _____ cm

   _____ in.

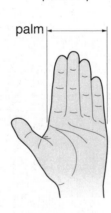

2. **Palm**

   _____ mm

   _____ cm

   _____ in.

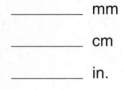

3. **Joint**

   _____ mm

   _____ cm

   _____ in.

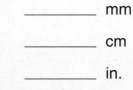

LESSON 6·2 **Personal Measures** *continued*

4. **Finger stretch**

   _____ mm

   _____ cm

   _____ in.

5. **Great span**

   _____ mm

   _____ cm

   _____ in.

6. **Cubit**

   _____ mm

   _____ cm

   _____ in.

7. **Fathom**

   _____ mm

   _____ cm

   _____ in.

8. **Natural yard**

   _____ mm

   _____ cm

   _____ in.

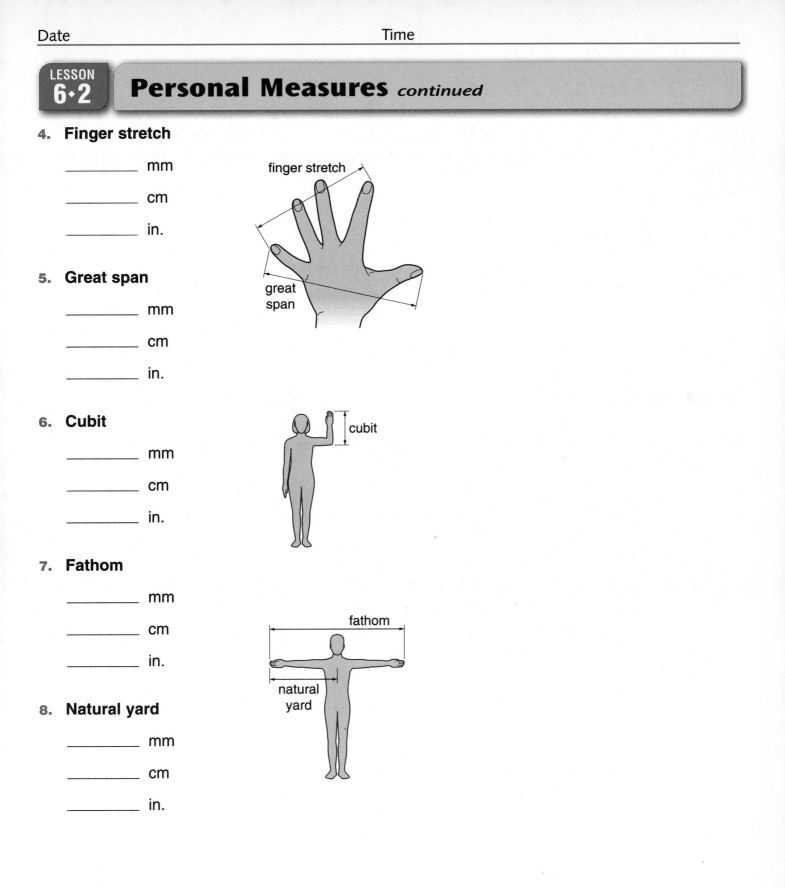

Finish First

Materials ☐ A deck of cards, consisting of four of each of the numbers 4, 5, 6, 7, and 8 (Do not use any other cards.)

Number of Players 2

Object of the Game To be the first to reach 21 or more

Directions

Decide who will go first. That person should always play first whenever you start a new game.

1. Shuffle the cards. Place the deck facedown.

2. The player going first turns over the top card and announces its value.

3. The player going second turns over the next card and announces the total value of the two cards turned over.

4. Partners continue to take turns turning over cards and announcing the total value of all the cards turned over so far.

5. The winner is the first player to correctly announce "21" or any number greater than 21.

6. Start a new game using the cards that are still facedown. If all the cards are turned over during a game, shuffle the deck, place it facedown, and continue.

LESSON 6·2   Estimation Challenge

A **fair game** is one in which each player has the same chance of winning. If there is an advantage or disadvantage in playing first, then the game is not fair.

With your partner, investigate whether *Finish First* is a fair game.

1. Collect data by playing the game. Over the next week, play *Finish First* at least 50 times. Keep a tally each day. Show how many times the player going first wins, and how many times the player going second wins.

Date	Player Going First Wins	Player Going Second Wins	Total Games to Date

2. Each day you play the game, record the results on the classroom tally sheet.

3. Each day you play, ask yourself:

   ◆ What is my estimate for the chance that the player going first will win?

   ◆ What is my estimate for the chance that the player going second will win?

   ◆ Do my estimates change as more and more games are played?

   ◆ Does *Finish First* seem to be a fair game?

LESSON 6·2   Math Boxes

1. Solve.   Solution

    a. $49 / e = 7$ _____

    b. $240 = 8 * t$ _____

    c. $r = 640 / 8$ _____

    d. $a = 187 - 38$ _____

    e. $c - 705 = 428$ _____

    SRB 218 219 222

2. Divide mentally.

    a. $829 / 4 \rightarrow$ _____

    b. $608 \div 3 \rightarrow$ _____

    c. $943 \div 2 \rightarrow$ _____

    d. $780 \div 5 \rightarrow$ _____

    e. $698 / 7 \rightarrow$ _____

    SRB 21

3. Write a 9-digit numeral with

3 in the hundredths place,
6 in the ten-thousands place,
4 in the thousandths place,
5 in the hundred-thousands place,
2 in the tens place, and
0 in all other places.

___ ___ ___, ___ ___ ___. ___ ___ ___

Write the numeral in words.

_____

_____

    SRB 4 28

4. Complete the table.

Standard Notation	Exponential Notation
10,000	
	10^3
	10^8
1,000,000,000	
	10^5

    SRB 5

5. A store is giving a 30% discount on all items. How much is saved on each item?

Regular Price	Savings
$35	
$246	
$3.50	
$78.50	

    SRB 51

6. Divora bakes a pie. She eats $\frac{1}{4}$ of the pie and her friend eats $\frac{1}{8}$. What fraction of the pie have they eaten? _____

How much is left? _____

    SRB 68

LESSON 6·3 Hand Measures: The Great Span

For measurements on this page and the next page:

If you are right-handed, measure your left hand.
If you are left-handed, measure your right hand.

Your **great span** is the distance from the tip of your thumb to the tip of your little finger. Place the tip of your thumb at the top of the ruler in the margin (at 0). Extend your fingers. Stretch your little finger as far along the ruler as you can. Read your great span measurement to the nearest millimeter, and record it below.

My great span is about _____ mm.

Your teacher will show you how to use the table below. Use it to record the great-span data for your class. The result is called a **stem-and-leaf plot.**

Great-Span Measurements for the Class (Millimeters)

Stems (100s and 10s)	Leaves (1s)
13	_____
14	_____
15	_____
16	_____
17	_____
18	_____
19	_____
20	_____
21	_____
22	_____
23	_____
24	_____

Landmarks for the Class Great-Span Data

minimum: _____ mm

maximum: _____ mm

mode(s): _____ mm

median: _____ mm

173

LESSON 6·3

Finger Measures: Finger Flexibility

The picture shows how to measure the **angle of separation** between your thumb and first (index) finger. This is a measure of finger flexibility.

1. Spread your thumb and first finger as far apart as you can. Do this in the air. Don't use your other hand to help. Lower your hand onto a sheet of paper. Trace around your thumb and first finger. With a straightedge, draw two line segments to make a V shape, or angle, that fits the finger opening. Use a protractor to measure the angle between your thumb and first finger. Record the measure of the angle.

Measure this angle.

Angle formed by thumb and first finger:

_____°

2. In the air, spread your first and second fingers as far apart as possible. On a sheet of paper, trace these fingers, and draw the angle of separation between them. Measure the angle and record its measure.

Angle formed by first and second fingers:

_____°

3. Record the class landmarks for both finger-separation angles in the table at the right.

Landmark	Thumb and First	First and Second
Minimum		
Maximum		
Mode(s)		
Median		

LESSON 6·3 Math Boxes

1. Fill in the missing numbers on the number lines.

a.

7 _____ 77

_____   _____   _____   _____

b.

0 _____ 68

_____   _____   _____

SRB
230–233

2. Make up a set of at least 12 numbers that have the following landmarks.

maximum: 73
minimum: 67
median: 70
mode: 70

_____

_____

Make a bar graph for this set of numbers.

SRB
119 122

3. Write five names for 100.

_____

_____

_____

_____

_____

SRB
219

4. Write each number in standard notation.

a. $8^2 =$ _____

b. $9^3 =$ _____

c. $5^3 =$ _____

d. $6 * 10^2 =$ _____

e. $9 * 10^3 =$ _____

SRB
6 8

175

LESSON 6·4   Mystery Plots

There are five line plots on page 177. Each plot shows a different set of data about a fifth-grade class.

Match each of the following four data set descriptions with one of the five plots. Then fill in the unit for each matched graph on page 177.

1. The number of hours of TV each fifth grader watched last night   Plot _____

2. The ages of the younger brothers and sisters of the fifth graders   Plot _____

3. The heights, in inches, of some fifth graders   Plot _____

4. The ages of some fifth graders' grandmothers   Plot _____

5. Explain how you selected the line plot for Data Set 4.

_____

_____

_____

_____

_____

6. Tell why you think the other line plots are not correct for Data Set 4.

_____

_____

_____

_____

_____

_____

LESSON 6·4

Mystery Plots *continued*

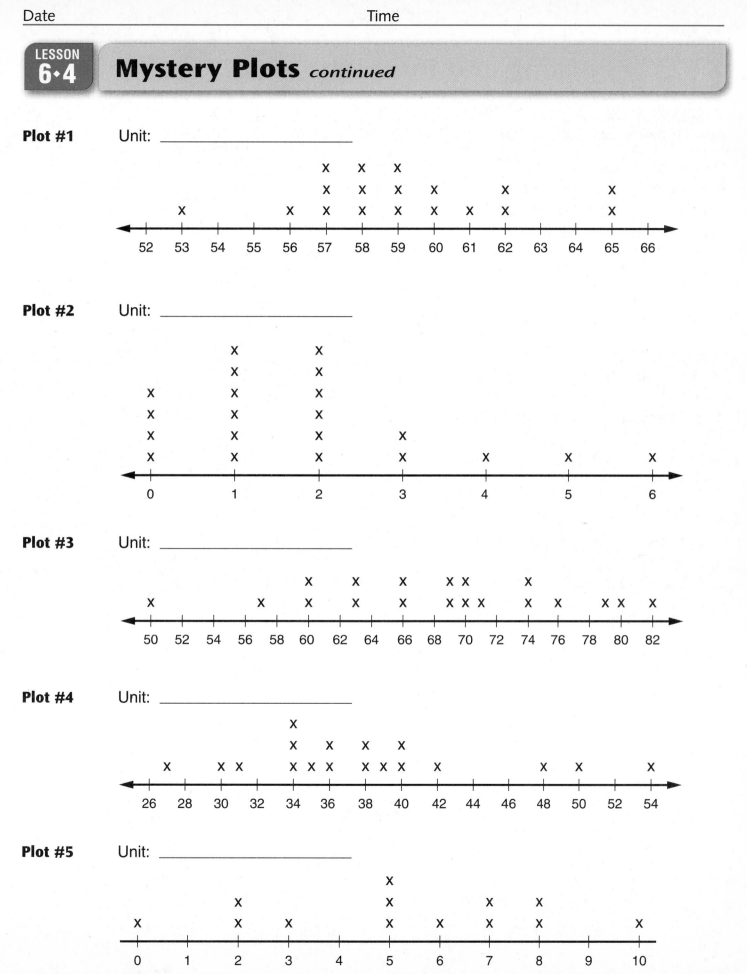

Plot #1   Unit: _____

Plot #2   Unit: _____

Plot #3   Unit: _____

Plot #4   Unit: _____

Plot #5   Unit: _____

Reaching and Jumping

Some fifth-grade students measured
how far they could reach and jump.

Each student stood with legs together,
feet flat on the floor, and one arm
stretched up as high as possible.
Arm reach was then measured
from top fingertip to floor.

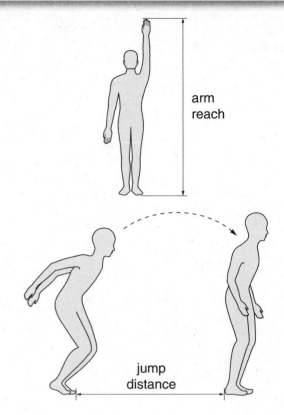

arm
reach

In the standing jump, each student
stood with knees bent and then
jumped forward as far as possible.
The distance was then measured from
the starting line to the point closest to
where the student's heels came down.

jump
distance

The students made stem-and-leaf plots of the results.

1. **a.** Which stem-and-leaf plot below shows arm reach?   Plot _____

   b. What is the median arm reach?   _____ in.

2. **a.** Which stem-and-leaf plot below shows standing-jump distances?   Plot _____

   b. What is the median standing-jump distance?   _____ in.

Plot #1
Unit: Inches

Stems (10s)	Leaves (1s)
4	4 6 8
5	0 0 3 3 4 5 6 7 7 8 8 9
6	0 0 1 3 3 8

Plot #2
Unit: Inches

Stems (10s)	Leaves (1s)
6	7
7	0 1 2 2 2 2 3 3 4 4 6 6 6 8 9 9
8	0 3 4 7

LESSON 6·4 Math Boxes

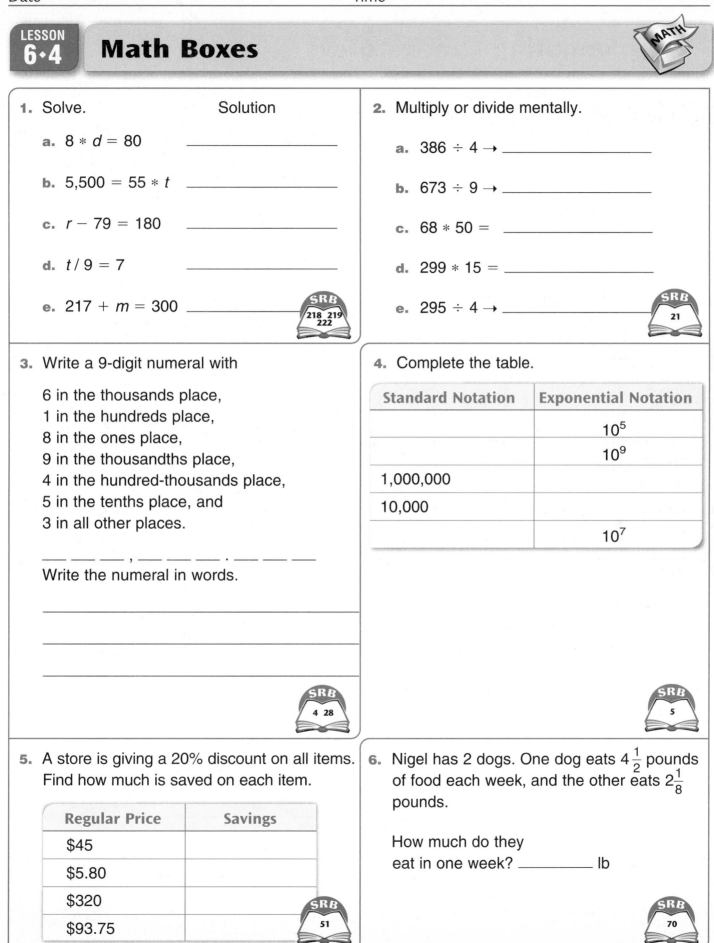

1. Solve.   Solution

   a. $8 * d = 80$ _____

   b. $5{,}500 = 55 * t$ _____

   c. $r - 79 = 180$ _____

   d. $t / 9 = 7$ _____

   e. $217 + m = 300$ _____

SRB 218 219 222

2. Multiply or divide mentally.

   a. $386 \div 4 \rightarrow$ _____

   b. $673 \div 9 \rightarrow$ _____

   c. $68 * 50 =$ _____

   d. $299 * 15 =$ _____

   e. $295 \div 4 \rightarrow$ _____

SRB 21

3. Write a 9-digit numeral with

   6 in the thousands place,
   1 in the hundreds place,
   8 in the ones place,
   9 in the thousandths place,
   4 in the hundred-thousands place,
   5 in the tenths place, and
   3 in all other places.

   ___ ___ ___ , ___ ___ ___ . ___ ___ ___

   Write the numeral in words.

   _____

   _____

   _____

SRB 4 28

4. Complete the table.

Standard Notation	Exponential Notation
	10^5
	10^9
1,000,000	
10,000	
	10^7

SRB 5

5. A store is giving a 20% discount on all items. Find how much is saved on each item.

Regular Price	Savings
$45	
$5.80	
$320	
$93.75	

SRB 51

6. Nigel has 2 dogs. One dog eats $4\frac{1}{2}$ pounds of food each week, and the other eats $2\frac{1}{8}$ pounds.

   How much do they
   eat in one week? _____ lb

SRB 70

Sampling Candy Colors

1. You and your partner each take 5 pieces of candy from the bowl. Combine your candies, and record your results in the table under Our Sample of 10 Candies.

Candy Color	Our Sample of 10 Candies		Combined Class Sample	
	Count	Percent	Count	Percent

2. Your class will work together to make a sample of 100 candies. Record the counts and percents of the class sample under Combined Class Sample in the table.

3. Finally, your class will count the total number of candies in the bowl and the number of each color.

   a. How well did your sample of 10 candies predict the number of each color in

   the bowl? _____

   b. How well did the combined class sample predict the number of each color in

   the bowl? _____

   c. Do you think that a larger sample is more trustworthy than a smaller

   sample? _____

   Explain your answer. _____

   _____

   _____

   _____

Math Boxes

---

1. Find the median and mean for each set of numbers.

a. 17, 13, 27, 33, 25

median: _____

mean: _____

b. 47, 29, 53, 46, 43, 32

median: _____

mean: _____

SRB 119–121

---

2. Estimate the sum.

3.1 + 0.72 + 34.7

Choose the best answer.

◯ 1,370 and 1,380

◯ 440 and 460

◯ 35 and 40

◯ 13 and 15

SRB 247

---

3. Draw a circle graph that is divided into the following sectors: 32%, 4%, 22%, 18%, and 24%. Make up a situation for the graph. Give the graph a title. Label each section.

Description:

_____

_____

_____

_____
(title)

SRB 125 126

---

4. Solve.

a. $34.2 \div 6 =$ _____

b. $39.05 / 5 =$ _____

c. $3\overline{)693.6}$

SRB 42 43

---

5. Write > or < to make the sentence true.

a. 662 _____ 626

b. 7,341 _____ 7,347

c. 203,467 _____ 203,764

d. 699,842 _____ 699,428

e. 521,369 _____ 531,399

SRB 9

LESSON 6·6   Is *Finish First* a Fair Game?

Math Message

1. What is the total number of *Finish First* games your class has played? _____ games

2. How many games did the player going first win? _____ games

3. How many games did the player going second win? _____ games

4. What is your best estimate for the chance the player going first will win? _____

5. What is your best estimate for the chance the player going second will win? _____

6. Did your estimates change as more games were played? _____

7. Is *Finish First* a fair game? _____

   Why or why not? _____

   _____

   _____

   _____

   _____

   _____

   If *Finish First* isn't a fair game, how could you make it more fair?

   _____

   _____

   _____

   _____

   _____

   _____

LESSON 6·6 Frequency Tables

A **frequency table** is a chart on which data is tallied to find the frequency of given events or values.

Use the frequency tables below to tally the Entertainment data and Favorite-Sports data on page 116 in your *Student Reference Book*. Then complete the tables. If you conducted your own survey, use the frequency tables to tally the data you collected. Then complete the tables.

1. What is the survey question? _____

_____

Category	Tallies	Number	Fraction	Percent

Total number of tallies _____

2. What is the survey question? _____

_____

Category	Tallies	Number	Fraction	Percent

Total number of tallies _____

LESSON 6·6

Data Graphs and Plots

1. Draw a bar graph for one of the survey questions on journal page 183. Label the parts of the graph. Give the graph a title.

(title)
_____

_____ _____ _____ _____ _____ _____

2. Draw a circle graph for the other survey question on journal page 183. Label the sections of the graph. Give the graph a title.

(title)
_____

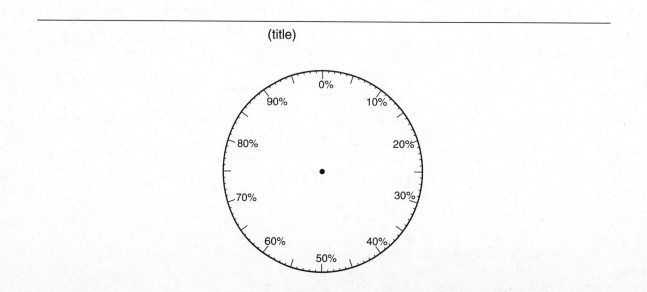

LESSON 6·6

Data Graphs and Plots *continued*

3. Make a stem-and-leaf plot for the Shower/Bath Time data on page 116 in your *Student Reference Book.* If you conducted your own survey, make a stem-and-leaf plot for the data you collected.

_____ _____

                                        (title)   (unit)

Stems (10s)	Leaves (1s)

4. Find the landmarks for this set of data.

   minimum: _____

   maximum: _____

   range: _____

   median: _____

   mode: _____

5. Calculate the mean (average).

   mean: _____

LESSON 6·6   Math Boxes

1. Write the correct fraction in each of the smaller regions of the figure below. Check to see that the fractional parts add up to 1.

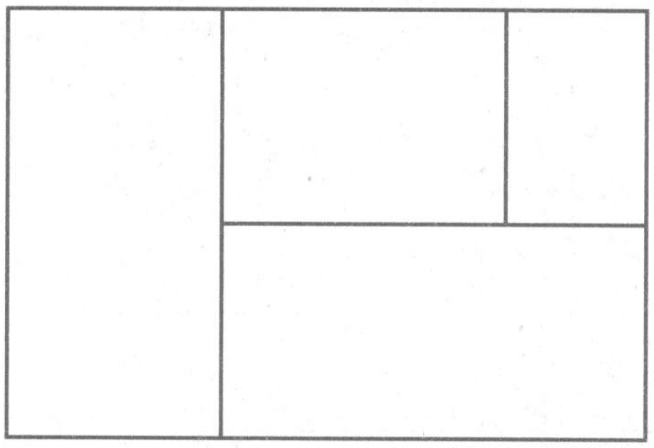

SRB 75

2.

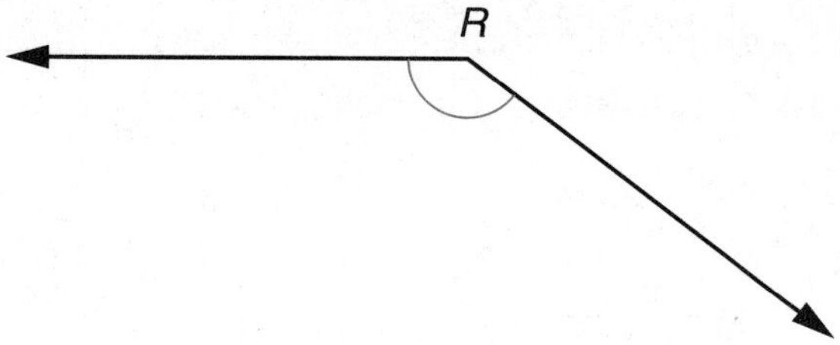

R

Estimate of the measure of ∠R: _____

The measure of ∠R is about _____.

SRB 204

3. Here are the results of Jay's last 11 science tests:

13, 3, 11, 19, 11, 16, 11, 17, 15, 12, 15

a. What is the mean of his scores? _____

b. What is the median of his scores? _____

c. What is the mode of his scores? _____

SRB 119–121

4. Measure each to the nearest centimeter.

a. *Student Reference Book*

length: _____ cm   width: _____ cm

b. seat of chair

length: _____ cm   width: _____ cm

c. sole of shoe   length: _____ cm

width: _____ cm

SRB 183

5. Measure each line segment to the nearest $\frac{1}{8}$ inch.

a. _____

_____

b. _____

_____

SRB 183

6. Write each fraction as a mixed number or a whole number.

a. $\frac{39}{4}$ = _____

b. $\frac{62}{7}$ = _____

c. $\frac{45}{6}$ = _____

d. $\frac{200}{5}$ = _____

e. $\frac{83}{9}$ = _____

SRB 62 63

LESSON 6·7 Climate Maps

To answer the questions below, use the Average Yearly Precipitation in the United States and Growing Seasons in the United States maps on page 380 of the American Tour section of your *Student Reference Book.*

The precipitation map shows the average amount of moisture that falls as sleet, hail, rain, or snow in one year. Snow is translated into an equivalent amount of rain.

The growing seasons map shows the average number of months between the last frost in spring and the first frost in fall. During this time, the temperature remains above freezing (32°F or 0°C), and crops can be grown.

1. Denver, Colorado, receives about _____ in. of precipitation as rain and snow per year.

   Denver's growing season is about _____ months long.

2. Los Angeles, California, receives about _____ in. of precipitation per year.

   The growing season in Los Angeles is _____ months long.

3. a. According to these maps, how are Los Angeles and New Orleans similar?

   _____

   _____

   _____

   b. Who is more likely to be worried about a lack of rain: a farmer near Los Angeles or a farmer near New Orleans? Why?

   _____

   _____

   _____

   _____

   _____

**LESSON
6·7**   **Climate Maps** *continued*

4. In general, does it rain more in the eastern states or in the western states?

   _____

5. In general, is the growing season longer in the northern states or in the southern states?

   _____

6. Cotton needs a growing season of at least 6 months. In the list below, circle the states most likely to grow cotton.

   Texas   Nebraska   Mississippi   Ohio

7. North Dakota and Kansas are the largest wheat-producing states.

   What is the length of the growing season in North Dakota? _____

   What is the length of the growing season in Kansas?   _____

   About how much precipitation does North Dakota
   receive per year?   _____

   About how much precipitation does eastern Kansas
   receive per year?   _____

8. a. Locate the Rocky Mountains on your landform map (American Tour, page 381).

      What is the growing season for this mountain area?

      _____

      _____

   b. What is the growing season for the Appalachian Mountains area?

      _____

      _____

Number Stories

1. Brenda bought 4 cheeseburgers for her family for lunch. The total cost was $5.56.

   How much did 2 cheeseburgers cost? _____

2. Thomas's family went on a long trip over summer vacation. They drove for 5 days. The distances for the 5 days were as follows: 347 miles, 504 miles, 393 miles, 422 miles, and 418 miles.

   a. To the nearest mile, what was the average distance traveled per day? _____

   b. Tell what you did with the remainder. Explain why. _____

   _____

   _____

3. Justin's school has 15 classrooms. On average, there are 28 students per room. One fifth of the classrooms are for fifth graders. About how many students are in the school?

   _____

   _____

4. Jamokas reads 45 pages of a book every night. How many pages did she read in

   the month of March (31 days)? _____

5. Lucienne and her class made 684 notecards for a school benefit.

   a. How many boxes of eight can they fill? _____

   b. Tell what you did with the remainder. Explain why.

   _____

   _____

   _____

   _____

LESSON 6·7 Math Boxes

1. Find the median and mean for each set of numbers.

    a. 33, 41, 37, 27, 32

        median: _____

        mean: _____

    b. 156, 102, 110, 200, 147

        median: _____

        mean: _____

SRB 119–121

2. $6.35 * 42$

Choose the best answer.
The product is between:

⬭   0.24 and 0.27

⬭   24 and 27

⬭   240 and 270

⬭   24,000 and 27,000

SRB 247

3. Ms. Allende's fifth graders collected information on favorite board games. Complete the table and make a circle graph of the data.

Favorite Game

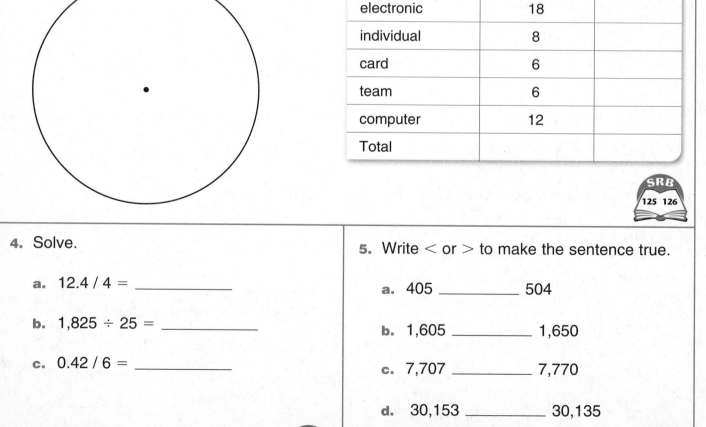

Favorite Games	Number of Students	Percent of Class
electronic	18	
individual	8	
card	6	
team	6	
computer	12	
Total		

SRB 125 126

4. Solve.

    a. $12.4 / 4 =$ _____

    b. $1,825 \div 25 =$ _____

    c. $0.42 / 6 =$ _____

SRB 42 43

5. Write $<$ or $>$ to make the sentence true.

    a. 405 _____ 504

    b. 1,605 _____ 1,650

    c. 7,707 _____ 7,770

    d. 30,153 _____ 30,135

    e. 697,707 _____ 697,077

SRB 9

LESSON 6·8   Calculating Fractions on a Slide Rule

Use your slide rule, or any other method, to add or subtract.

1. $\frac{1}{2} + \frac{1}{4} =$ _____

2. $\frac{5}{8} + \frac{2}{8} =$ _____

3. $2\frac{1}{2} + 3 =$ _____

4. $3\frac{5}{8} + 3\frac{3}{4} =$ _____

5. $1\frac{9}{16} + 1\frac{5}{16} =$ _____

6. $\frac{7}{8} - \frac{3}{8} =$ _____

7. $5\frac{3}{4} - 2\frac{1}{4} =$ _____

8. $7\frac{1}{2} - 4\frac{5}{8} =$ _____

9. $\frac{19}{16} - \frac{1}{2} =$ _____

10. $5\frac{1}{2} - 6 =$ _____

11. Put a star next to the above problems that you thought were the easiest.

12. Complete the following:

It is easy to add or subtract fractions with the same denominator (for example, $\frac{4}{8} - \frac{3}{8}$)

because... _____

_____

_____

_____

Try This

13. Melanie and Jamal are making posters for their classroom.

Melanie's poster shows addition of fractions as: $\frac{a}{c} + \frac{b}{c} = \frac{a+b}{c}$

Jamal's poster shows subtraction of fractions as: $\frac{a}{c} - \frac{b}{c} = \frac{a}{b} - c$

Explain how they can check that their posters are correct?

_____

_____

_____

_____

_____

LESSON 6·8   Calculating with Fraction Sticks

Write the missing fraction for each pair of fraction sticks. Then write
the sum or difference of the fractions.

1. $\frac{5}{12}$ + _____ = _____

2. $\frac{5}{6}$ − _____ = _____

3. _____ − $\frac{1}{4}$ = _____

4. $\frac{3}{5}$ + _____ = _____

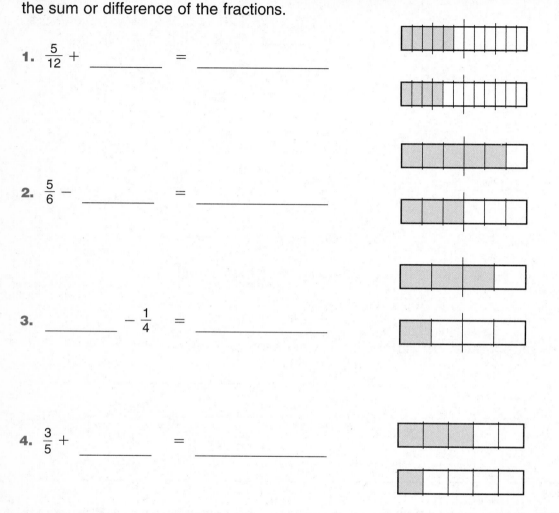

5. Andy jogs on a track where each lap is $\frac{1}{4}$ mile. Find the number of miles he jogged
each day and then the total number of laps and miles for the three days.

Day	Laps	Distance
Monday	5	
Wednesday	10	
Thursday	8	
Total		

LESSON 6·8

Math Boxes

1. **a.** What is 30% of $1.00?

   _____

   b. How many cents is 60% of $1.00?

   _____

   c. What is 75% of $1.00?

   _____

   d. How many cents is 5% of $1.00?

   _____

   SRB 51

2. Use a calculator to complete the table. (Round decimals to the nearest hundredth.)

Fraction	Decimal	Percent
$\frac{3}{7}$		
$\frac{10}{11}$		
$\frac{8}{15}$		
$\frac{7}{9}$		
$\frac{8}{14}$		

   SRB 89 90

3. Use your calculator to complete the table.

Exponential Notation	Product of Factors	Standard Notation
9^4		
	12 * 12 * 12 * 12	20,736
8^4		
	11 * 11 * 11 * 11 * 11	
10^3		

   SRB 5 6

4. Complete the "What's My Rule?" table and state the rule.

in	out
27	20
	6
5	−2
10	

   Rule

   _____

   SRB 231 232

5. Make true sentences by inserting parentheses.

   a. 5 * 4 − 2 = 10

   b. 25 + 8 * 7 = 81

   c. 36 / 6 − 5 = 36

   d. 45 / 9 + 6 = 11

   e. 45 / 9 + 6 = 3

   SRB 222 223

LESSON 6·9

Clock Fractions

Part 1: Math Message

The numbers on a clock face divide one hour into twelfths. Each $\frac{1}{12}$ of an hour is 5 minutes.

Whole
hour

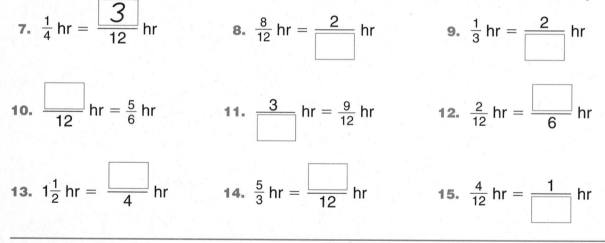

How many minutes does each of the following fractions and mixed numbers represent? The first one has been done for you.

1. $\frac{1}{12}$ hr = ___5___ min

2. $\frac{5}{12}$ hr = _____ min

3. $\frac{1}{2}$ hr = _____ min

4. $\frac{1}{3}$ hr = _____ min

5. $\frac{1}{4}$ hr = _____ min

6. $\frac{1}{6}$ hr = _____ min

Part 2

Using the clock face, fill in the missing numbers. The first one has been done for you.

7. $\frac{1}{4}$ hr = $\frac{3}{12}$ hr

8. $\frac{8}{12}$ hr = $\frac{2}{\square}$ hr

9. $\frac{1}{3}$ hr = $\frac{2}{\square}$ hr

10. $\frac{\square}{12}$ hr = $\frac{5}{6}$ hr

11. $\frac{3}{\square}$ hr = $\frac{9}{12}$ hr

12. $\frac{2}{12}$ hr = $\frac{\square}{6}$ hr

13. $1\frac{1}{2}$ hr = $\frac{\square}{4}$ hr

14. $\frac{5}{3}$ hr = $\frac{\square}{12}$ hr

15. $\frac{4}{12}$ hr = $\frac{1}{\square}$ hr

Part 3

Use clock fractions, if helpful, to solve these problems. Write each answer as a fraction.

Example: $\frac{3}{4} - \frac{1}{3} = ?$

Think: 45 minutes − 20 minutes = 25 minutes

So $\frac{3}{4} - \frac{1}{3} = \frac{5}{12}$

16. $\frac{5}{12} + \frac{3}{12} =$ _____

17. $\frac{3}{4} + \frac{2}{4} =$ _____

18. $\frac{11}{12} - \frac{3}{12} =$ _____

19. $1 - \frac{2}{3} =$ _____

20. $\frac{5}{4} - \frac{2}{4} =$ _____

21. $\frac{2}{3} + \frac{1}{6} =$ _____

22. $\frac{1}{4} + \frac{1}{3} =$ _____

23. $\frac{1}{3} - \frac{1}{4} =$ _____

24. $\frac{5}{6} - \frac{3}{4} =$ _____

LESSON 6·9 Number Strip Fractions

Name the strips that you used for the numerator and denominator. Then list the fractions formed by the two strips.

Problem 1

	Strip Name	Fractions List
Numerator:	_____	
Denominator:	_____	_____

Problem 2

	Strip Name	Fractions List
Numerator:	_____	
Denominator:	_____	_____

Problem 3

	Strip Name	Fractions List
Numerator:	_____	
Denominator:	_____	_____

Problem 4

Explain how you can use a multiplication table to find equivalent fractions for $\frac{9}{27}$.

_____

_____

_____

_____

LESSON 6·9 — Using a Common Denominator

Study the examples. Then work the problems below in the same way.

Example 1: $\frac{2}{3} + \frac{1}{6} = ?$

Unlike Denominators		Common Denominators
$\frac{2}{3}$	$\frac{2}{3} = \frac{4}{6}$	$\frac{4}{6}$
$+ \frac{1}{6}$		$+ \frac{1}{6}$
		$\frac{5}{6}$

Example 2: $\frac{5}{6} - \frac{3}{4} = ?$

Unlike Denominators		Common Denominators
$\frac{5}{6}$	$\frac{5}{6} = \frac{10}{12}$	$\frac{10}{12}$
$- \frac{3}{4}$	$\frac{3}{4} = \frac{9}{12}$	$- \frac{9}{12}$
		$\frac{1}{12}$

1. $\frac{2}{3} + \frac{2}{9} = ?$

Unlike Denominators	Common Denominators
$\frac{2}{3}$	
$+ \frac{2}{9}$	

2. $\frac{13}{16} - \frac{3}{4} = ?$

Unlike Denominators	Common Denominators
$\frac{13}{16}$	
$- \frac{3}{4}$	

3. $\frac{1}{3} + \frac{2}{5} = ?$

Unlike Denominators	Common Denominators
$\frac{1}{3}$	
$+ \frac{2}{5}$	

4. $\frac{5}{6} - \frac{4}{9} = ?$

Unlike Denominators	Common Denominators
$\frac{5}{6}$	
$- \frac{4}{9}$	

LESSON 6·9

Using a Common Denominator *continued*

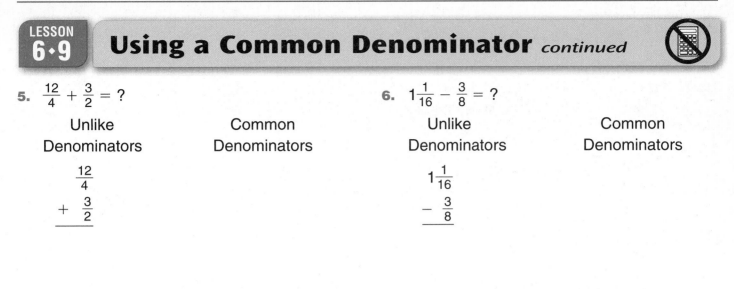

5. $\frac{12}{4} + \frac{3}{2} = ?$

Unlike Denominators	Common Denominators

$$\begin{array}{r} \frac{12}{4} \\ + \ \frac{3}{2} \\ \hline \end{array}$$

6. $1\frac{1}{16} - \frac{3}{8} = ?$

Unlike Denominators	Common Denominators

$$\begin{array}{r} 1\frac{1}{16} \\ - \ \frac{3}{8} \\ \hline \end{array}$$

7. A piece of ribbon is $7\frac{1}{2}$ in. long. If a piece $2\frac{3}{16}$ in. long is cut off, how long is the remaining piece? _____ in.

Write a number sentence to show how you solved the problem.

_____

8. Three boards are glued together. The diagram below shows the thickness of each board. What is the total thickness of the three boards? _____ in.

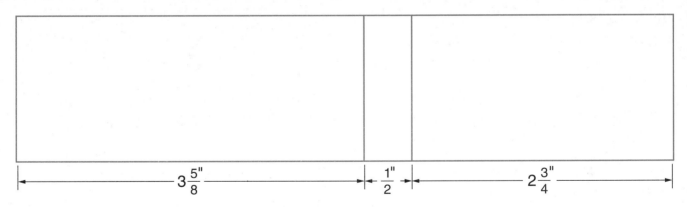

Write a number sentence to show how you solved the problem.

_____

LESSON 6·9 *Fraction Capture*

Materials
☐ *Fraction Capture* gameboard
☐ 2 six-sided dice

Players
2

Object of the Game
To capture the most squares on the *Fraction Capture* gameboard. A player captures a square if he or she shades more than $\frac{1}{2}$ of it.

Directions

1. Player 1 rolls two dice and makes a fraction with the numbers that come up. The number on either die can be the denominator. The number on the other die becomes the numerator.
   A fraction equal to a whole number is NOT allowed. For example, if a player rolls 3 and 6, the fraction can't be $\frac{6}{3}$, because $\frac{6}{3}$ equals 2.

2. Player 1 initials sections of one or more gameboard squares to show the fraction formed. This claims the sections for the player.

   Example: The player rolls a 4 and 5 and makes $\frac{5}{4}$. The player claims five $\frac{1}{4}$ sections by initialing them.

   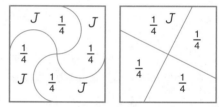

   ◆ Equivalent fractions can be claimed. For example, if a player rolls 1 and 2 and makes $\frac{1}{2}$, the player can initial one $\frac{1}{2}$ section of a square, or two $\frac{1}{4}$ sections, or three $\frac{1}{6}$ sections.

   ◆ The fraction may be split between squares. For example, a player can show $\frac{4}{3}$ by claiming $\frac{2}{3}$ on one square and $\frac{2}{3}$ on another square. However, all of the fractions must be shown.

3. Players take turns. If a player can't form a fraction and claim enough sections to show that fraction, the player's turn is over.

4. A player captures a square when that player has claimed sections making up more than $\frac{1}{2}$ of the square. If each player has initialed $\frac{1}{2}$ of a square, no one has captured that square.

   ◆ Blocking is allowed. For example, if Player 1 initials $\frac{1}{2}$ of a square, Player 2 may initial the other half so no one can capture the square.

5. Play ends when all of the squares have been captured or blocked. The winner is the player who has captured the most squares.

LESSON 6·9 Math Boxes

1. In the figure below, write the correct fraction in each of the smaller regions. Check to see that the fractional parts add up to 1.

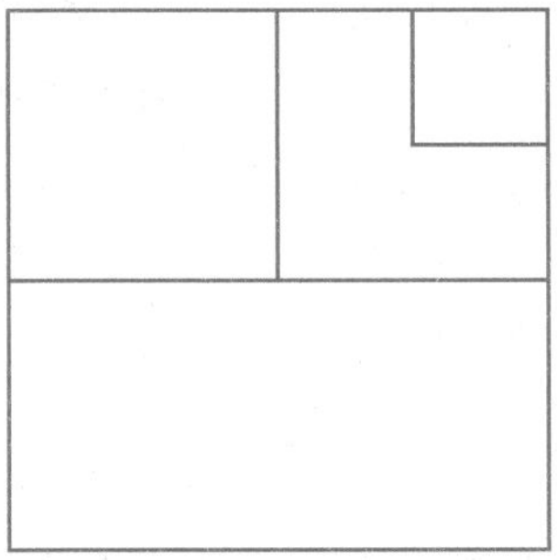

SRB
75

2.

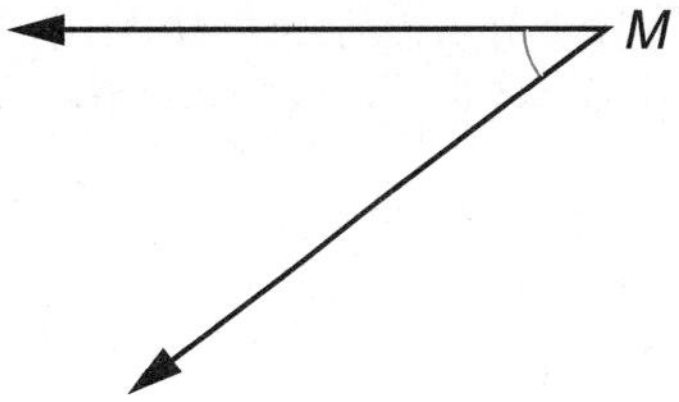

Estimate the measure of $\angle M$: _____

The measure of $\angle M$ is about _____.

SRB
204

3. Esther did 5 standing jumps. Her longest jump was 50 in. Could her average jump be 53 in.?

_____

Create a data set for Esther's jumps that could have this average.

_____

SRB
121

4. Measure the length and width of each of the following objects to the nearest centimeter.

a. pinkie finger   b. pencil

length: _____ cm   length: _____ cm

width: _____ cm   width: _____ cm

c. notebook

length: _____ cm

width: _____ cm

SRB
183

5. Measure each line segment to the nearest $\frac{1}{8}$ in.

a. _____

_____ in.

b. _____

_____ in.

SRB
183

6. Rename each fraction as a mixed number or a whole number.

a. $\frac{59}{5} = $ _____

b. $\frac{88}{11} = $ _____

c. $\frac{120}{7} = $ _____

d. $\frac{94}{4} = $ _____

e. $\frac{102}{6} = $ _____

SRB
62 63

LESSON 6·10   Finding Common Denominators

1. a. Draw a horizontal line to split each part of this thirds fraction stick into 2 equal parts. How many parts are there in all? _____

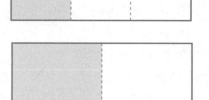

b. Draw horizontal lines to split each part of this halves fraction stick into 3 equal parts. How many parts are there in all? _____

c. $\dfrac{\boxed{} * 1}{\boxed{} * 3} = \dfrac{2}{6}$ $\dfrac{\boxed{} * 1}{\boxed{} * 2} = \dfrac{3}{6}$

2. a. If you drew lines to split each part of this fourths fraction stick into 6 equal parts, how many parts would there be in all? _____

b. If you drew lines to split each part of this sixths fraction stick into 4 equal parts, how many parts would there be in all? _____

c. $\dfrac{\boxed{} * 3}{\boxed{} * 4} = \dfrac{18}{24}$ $\dfrac{\boxed{} * 5}{\boxed{} * 6} = \dfrac{20}{24}$

3. One way to find a common denominator for a pair of fractions is to make a list of equivalent fractions.

$$\frac{3}{4} = \frac{6}{8} = \frac{9}{12} = \frac{12}{16} = \frac{15}{20} = \frac{18}{24} = \cdots \qquad \frac{5}{6} = \frac{10}{12} = \frac{15}{18} = \frac{20}{24} = \cdots$$

Another way to find a common denominator for a pair of fractions is ...

_____

Try This

Name a common denominator for each pair of fractions.

4. $\dfrac{3}{4}$ and $\dfrac{5}{16}$ = _____

5. $\dfrac{5}{8}$ and $\dfrac{9}{10}$ = _____

6. $\dfrac{4}{5}$ and $\dfrac{5}{6}$ = _____

Give the values of the variables that make each equation true.

7. $\dfrac{t * 4}{t * 7} = \dfrac{12}{21}$

$t =$ _____

8. $\dfrac{m * 4}{m * 6} = \dfrac{n}{30}$

$m =$ _____ $n =$ _____

9. $\dfrac{8 * x}{5 * x} = \dfrac{y}{45}$

$x =$ _____ $y =$ _____

LESSON 6·10 Using a Common Denominator

Common denominators are useful not only for adding and subtracting fractions, but also for comparing fractions.

A quick way to find a common denominator for a pair of fractions is to find the product of the denominators.

Example: Compare $\frac{2}{3}$ and $\frac{5}{8}$. Use 3 * 8 as a common denominator.

$$\frac{2}{3} = \frac{(8 * 2)}{(8 * 3)} = \frac{16}{24} \qquad \frac{5}{8} = \frac{(3 * 5)}{(3 * 8)} = \frac{15}{24}$$

$$\frac{16}{24} > \frac{15}{24}, \text{ so } \frac{2}{3} > \frac{5}{8}.$$

1. Rewrite each pair of fractions below as equivalent fractions with a common denominator. Then write < (less than) or > (greater than) to compare the fractions.

	Original Fraction	Equivalent Fraction	> or <
a.	$\frac{4}{7}$		$\frac{4}{7}$ $\frac{3}{5}$
	$\frac{3}{5}$		
b.	$\frac{9}{4}$		$\frac{9}{4}$ $\frac{7}{3}$
	$\frac{7}{3}$		

Find a common denominator. Then add or subtract.

2. $\frac{1}{2} - \frac{1}{3} =$ _____

3. $\frac{7}{8} + \frac{2}{5} =$ _____

4. $\frac{3}{4} - \frac{1}{2} =$ _____

5. $\frac{4}{5} + \frac{2}{3} =$ _____

6. $\begin{array}{r} \frac{9}{10} \\ - \frac{5}{6} \\ \hline \end{array}$

7. $\begin{array}{r} \frac{1}{10} \\ + \frac{3}{4} \\ \hline \end{array}$

LESSON 6·10 Fraction Problems

1. To maintain their energy during the racing season, professional bicycle racers eat between 6,000 and 8,000 calories per day.

   About $\frac{3}{20}$ of these calories come from fat, and about $\frac{5}{20}$ come from protein. The remaining calories come from carbohydrates.

   What fraction of a bicycle racer's calories comes from carbohydrates? _____

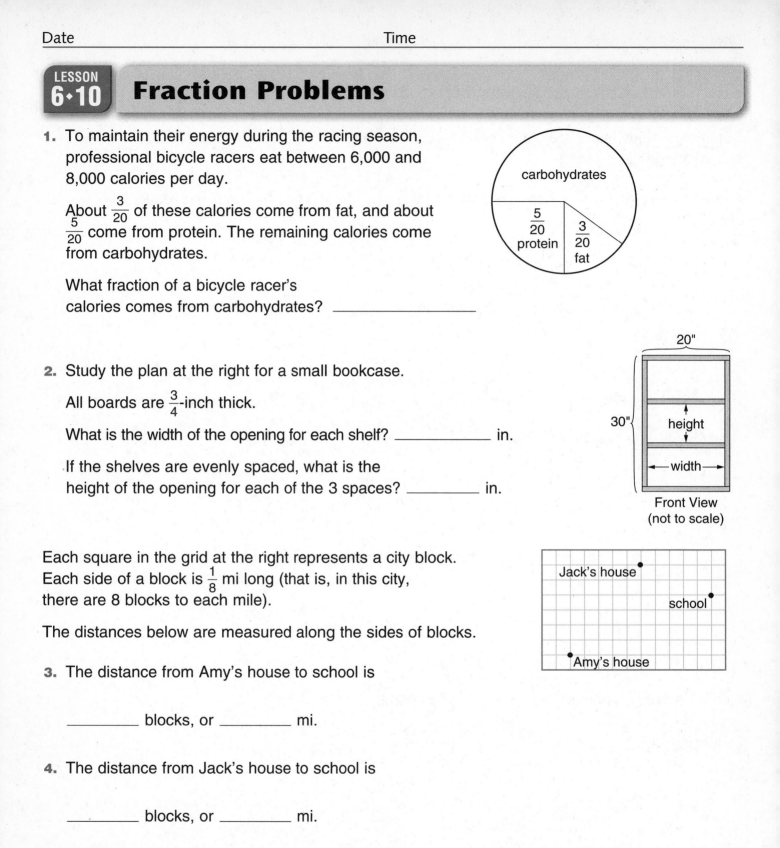

carbohydrates

$\frac{5}{20}$ protein $\frac{3}{20}$ fat

2. Study the plan at the right for a small bookcase.

   All boards are $\frac{3}{4}$-inch thick.

   What is the width of the opening for each shelf? _____ in.

   If the shelves are evenly spaced, what is the height of the opening for each of the 3 spaces? _____ in.

20"

30"   height

width

Front View
(not to scale)

Each square in the grid at the right represents a city block. Each side of a block is $\frac{1}{8}$ mi long (that is, in this city, there are 8 blocks to each mile).

The distances below are measured along the sides of blocks.

Jack's house

school

Amy's house

3. The distance from Amy's house to school is

   _____ blocks, or _____ mi.

4. The distance from Jack's house to school is

   _____ blocks, or _____ mi.

5. How much farther from school is Amy's house than Jack's house? _____ mi

6. Amy walks from school to Jack's house and then home.

   How far is that? _____ mi

LESSON 6·10   **Math Boxes**

---

1. a. What is 50% of $10.00?

_____

b. What percent
of $10.00 is $7.50? _____

c. What is 35%
of $10.00? _____

d. What percent of
$10.00 is $3.20? _____

SRB 51

---

2. Use a calculator to complete the table.
(Round decimals to the nearest hundredth.)

Fraction	Decimal	Percent
$\frac{11}{12}$		
$\frac{5}{7}$		
$\frac{14}{15}$		
$\frac{5}{6}$		
$\frac{2}{9}$		

SRB 89 99

---

3. Use your calculator to complete the table.

Exponential Notation	Product of Factors	Standard Notation
4^4		
	$5 * 5 * 5$	
6^4		1,296
	$7 * 7 * 7 * 7 * 7$	
9^3		

SRB 5 6

---

4. Complete the "What's My Rule?"
table and state the rule.

Rule
_____

in	out
8	17
11	
5	14
	4

SRB 231 232

---

5. Make these sentences true by inserting
parentheses.

a. $19 + 41 * 3 = 180$

b. $5 = 16 / 2 + 2 - 5$

c. $-1 = 16 / 2 + 2 - 5$

d. $24 \div 8 + 4 * 3 = 6$

e. $24 \div 8 + 4 * 3 = 15$

SRB 222 223

LESSON 6·11   Math Boxes

1. Use your calculator to rename each of the following in standard notation.

a. $5^5 = $ _____

b. $7^3 = $ _____

c. $9^3 = $ _____

d. $3^9 = $ _____

SRB 6

2. Continue each pattern.

SRB 230

a. 5, 14, 23, _____, _____, _____

b. 48, _____, 152, 204, _____, _____

c. 12, 72, _____, 192, _____, _____

d. 60.5, 111, _____, _____, 262.5

e. _____, 21.50, 22.25, _____, 23.75

3. Write the following numbers in order from greatest to least.

50,200; 5,200; 52,000; 5,200,002; 500,200

_____

_____

_____

_____

_____

SRB 9

4. Complete the "What's My Rule?" table and state the rule.

in	out
20	800
3	120
40	
	2,000
	320
700	

Rule
_____

SRB 231 232

5. Write the mixed number name and fraction name by each diagram.

Whole
hexagon

a.

Mixed number _____

Fraction _____

b.

Mixed number _____

Fraction _____

SRB 62

6. Make true sentences by inserting parentheses.

a. $4 * 8 - 5 = 12$

b. $2 + 7 * 7 = 51$

c. $91 / 4 - 3 = 91$

d. $63 / 9 + 11 = 18$

e. $72 / 30 + 6 = 2$

SRB 222 223

Reference

Place-Value Chart

trillions	100B	10B	billions	100M	10M	millions	hundred-thousands	ten-thousands	thousands	hundreds	tens	ones	.	tenths	hundredths	thousandths
1,000 billions			1,000 millions			1,000,000s	100,000s	10,000s	1,000s	100s	10s	1s	.	0.1s	0.01s	0.001s
10^{12}	10^{11}	10^{10}	10^9	10^8	10^7	10^6	10^5	10^4	10^3	10^2	10^1	10^0	.	10^{-1}	10^{-2}	10^{-3}

Probability Meter

CERTAIN

Percent	Decimal	Fraction	
100%	1.00 / 0.99	$\frac{99}{100}$	1
95%	0.95	$\frac{19}{20}$	
90%	0.90	$\frac{9}{10}$	
	0.875	$\frac{7}{8}$	
85%	0.85	$\frac{5}{6}$	
	$0.8\overline{3}$		
80%	0.80	$\frac{4}{5}, \frac{8}{10}$	
75%	0.75	$\frac{3}{4}, \frac{6}{8}$	
70%	0.70	$\frac{7}{10}$	
	$0.6\overline{6}$	$\frac{2}{3}$	
65%	0.65		
	0.625	$\frac{5}{8}$	
60%	0.60	$\frac{3}{5}, \frac{6}{10}$	
55%	0.55		
50%	0.50	$\frac{1}{2}, \frac{2}{4}, \frac{3}{6}, \frac{4}{8}, \frac{5}{10}, \frac{10}{20}, \frac{50}{100}$	
45%	0.45		
40%	0.40	$\frac{2}{5}, \frac{4}{10}$	
	0.375	$\frac{3}{8}$	
35%	0.35		
	$0.3\overline{3}$	$\frac{1}{3}$	
30%	0.30	$\frac{3}{10}$	
25%	0.25	$\frac{1}{4}, \frac{2}{8}$	
20%	0.20	$\frac{1}{5}$	
	$0.1\overline{6}$	$\frac{1}{6}$	
15%	0.15		
	0.125	$\frac{1}{8}$	
10%	0.10	$\frac{1}{10}$	
5%	0.05	$\frac{1}{20}$	
0%	0.01 / 0.00	$\frac{1}{100}$	0

EXTREMELY LIKELY · VERY LIKELY · LIKELY · 50–50 CHANCE · UNLIKELY · VERY UNLIKELY · EXTREMELY UNLIKELY

IMPOSSIBLE

Symbols

+	plus or positive
−	minus or negative
*, ×	multiplied by
÷, /	divided by
=	is equal to
≠	is not equal to
<	is less than
>	is greater than
≤	is less than or equal to
≥	is greater than or equal to
x^n	nth power of x
$\sqrt{x}$	square root of x
%	percent
$a{:}b,\ a/b,\ \frac{a}{b}$	ratio of a to b or a divided by b or the fraction $\frac{a}{b}$
°	degree
(a,b)	ordered pair
$\overleftrightarrow{AS}$	line AS
$\overline{AS}$	line segment AS
$\overrightarrow{AS}$	ray AS
∟	right angle
⊥	is perpendicular to
‖	is parallel to
△ABC	triangle ABC
∠ABC	angle ABC
∠B	angle B

Reference

Equivalent Fractions, Decimals, and Percents

															Decimal	Percent
$\frac{1}{2}$	$\frac{2}{4}$	$\frac{3}{6}$	$\frac{4}{8}$	$\frac{5}{10}$	$\frac{6}{12}$	$\frac{7}{14}$	$\frac{8}{16}$	$\frac{9}{18}$	$\frac{10}{20}$	$\frac{11}{22}$	$\frac{12}{24}$	$\frac{13}{26}$	$\frac{14}{28}$	$\frac{15}{30}$	0.5	50%
$\frac{1}{3}$	$\frac{2}{6}$	$\frac{3}{9}$	$\frac{4}{12}$	$\frac{5}{15}$	$\frac{6}{18}$	$\frac{7}{21}$	$\frac{8}{24}$	$\frac{9}{27}$	$\frac{10}{30}$	$\frac{11}{33}$	$\frac{12}{36}$	$\frac{13}{39}$	$\frac{14}{42}$	$\frac{15}{45}$	$0.\overline{3}$	$33\frac{1}{3}\%$
$\frac{2}{3}$	$\frac{4}{6}$	$\frac{6}{9}$	$\frac{8}{12}$	$\frac{10}{15}$	$\frac{12}{18}$	$\frac{14}{21}$	$\frac{16}{24}$	$\frac{18}{27}$	$\frac{20}{30}$	$\frac{22}{33}$	$\frac{24}{36}$	$\frac{26}{39}$	$\frac{28}{42}$	$\frac{30}{45}$	$0.\overline{6}$	$66\frac{2}{3}\%$
$\frac{1}{4}$	$\frac{2}{8}$	$\frac{3}{12}$	$\frac{4}{16}$	$\frac{5}{20}$	$\frac{6}{24}$	$\frac{7}{28}$	$\frac{8}{32}$	$\frac{9}{36}$	$\frac{10}{40}$	$\frac{11}{44}$	$\frac{12}{48}$	$\frac{13}{52}$	$\frac{14}{56}$	$\frac{15}{60}$	0.25	25%
$\frac{3}{4}$	$\frac{6}{8}$	$\frac{9}{12}$	$\frac{12}{16}$	$\frac{15}{20}$	$\frac{18}{24}$	$\frac{21}{28}$	$\frac{24}{32}$	$\frac{27}{36}$	$\frac{30}{40}$	$\frac{33}{44}$	$\frac{36}{48}$	$\frac{39}{52}$	$\frac{42}{56}$	$\frac{45}{60}$	0.75	75%
$\frac{1}{5}$	$\frac{2}{10}$	$\frac{3}{15}$	$\frac{4}{20}$	$\frac{5}{25}$	$\frac{6}{30}$	$\frac{7}{35}$	$\frac{8}{40}$	$\frac{9}{45}$	$\frac{10}{50}$	$\frac{11}{55}$	$\frac{12}{60}$	$\frac{13}{65}$	$\frac{14}{70}$	$\frac{15}{75}$	0.2	20%
$\frac{2}{5}$	$\frac{4}{10}$	$\frac{6}{15}$	$\frac{8}{20}$	$\frac{10}{25}$	$\frac{12}{30}$	$\frac{14}{35}$	$\frac{16}{40}$	$\frac{18}{45}$	$\frac{20}{50}$	$\frac{22}{55}$	$\frac{24}{60}$	$\frac{26}{65}$	$\frac{28}{70}$	$\frac{30}{75}$	0.4	40%
$\frac{3}{5}$	$\frac{6}{10}$	$\frac{9}{15}$	$\frac{12}{20}$	$\frac{15}{25}$	$\frac{18}{30}$	$\frac{21}{35}$	$\frac{24}{40}$	$\frac{27}{45}$	$\frac{30}{50}$	$\frac{33}{55}$	$\frac{36}{60}$	$\frac{39}{65}$	$\frac{42}{70}$	$\frac{45}{75}$	0.6	60%
$\frac{4}{5}$	$\frac{8}{10}$	$\frac{12}{15}$	$\frac{16}{20}$	$\frac{20}{25}$	$\frac{24}{30}$	$\frac{28}{35}$	$\frac{32}{40}$	$\frac{36}{45}$	$\frac{40}{50}$	$\frac{44}{55}$	$\frac{48}{60}$	$\frac{52}{65}$	$\frac{56}{70}$	$\frac{60}{75}$	0.8	80%
$\frac{1}{6}$	$\frac{2}{12}$	$\frac{3}{18}$	$\frac{4}{24}$	$\frac{5}{30}$	$\frac{6}{36}$	$\frac{7}{42}$	$\frac{8}{48}$	$\frac{9}{54}$	$\frac{10}{60}$	$\frac{11}{66}$	$\frac{12}{72}$	$\frac{13}{78}$	$\frac{14}{84}$	$\frac{15}{90}$	$0.1\overline{6}$	$16\frac{2}{3}\%$
$\frac{5}{6}$	$\frac{10}{12}$	$\frac{15}{18}$	$\frac{20}{24}$	$\frac{25}{30}$	$\frac{30}{36}$	$\frac{35}{42}$	$\frac{40}{48}$	$\frac{45}{54}$	$\frac{50}{60}$	$\frac{55}{66}$	$\frac{60}{72}$	$\frac{65}{78}$	$\frac{70}{84}$	$\frac{75}{90}$	$0.8\overline{3}$	$83\frac{1}{3}\%$
$\frac{1}{7}$	$\frac{2}{14}$	$\frac{3}{21}$	$\frac{4}{28}$	$\frac{5}{35}$	$\frac{6}{42}$	$\frac{7}{49}$	$\frac{8}{56}$	$\frac{9}{63}$	$\frac{10}{70}$	$\frac{11}{77}$	$\frac{12}{84}$	$\frac{13}{91}$	$\frac{14}{98}$	$\frac{15}{105}$	0.143	14.3%
$\frac{2}{7}$	$\frac{4}{14}$	$\frac{6}{21}$	$\frac{8}{28}$	$\frac{10}{35}$	$\frac{12}{42}$	$\frac{14}{49}$	$\frac{16}{56}$	$\frac{18}{63}$	$\frac{20}{70}$	$\frac{22}{77}$	$\frac{24}{84}$	$\frac{26}{91}$	$\frac{28}{98}$	$\frac{30}{105}$	0.286	28.6%
$\frac{3}{7}$	$\frac{6}{14}$	$\frac{9}{21}$	$\frac{12}{28}$	$\frac{15}{35}$	$\frac{18}{42}$	$\frac{21}{49}$	$\frac{24}{56}$	$\frac{27}{63}$	$\frac{30}{70}$	$\frac{33}{77}$	$\frac{36}{84}$	$\frac{39}{91}$	$\frac{42}{98}$	$\frac{45}{105}$	0.429	42.9%
$\frac{4}{7}$	$\frac{8}{14}$	$\frac{12}{21}$	$\frac{16}{28}$	$\frac{20}{35}$	$\frac{24}{42}$	$\frac{28}{49}$	$\frac{32}{56}$	$\frac{36}{63}$	$\frac{40}{70}$	$\frac{44}{77}$	$\frac{48}{84}$	$\frac{52}{91}$	$\frac{56}{98}$	$\frac{60}{105}$	0.571	57.1%
$\frac{5}{7}$	$\frac{10}{14}$	$\frac{15}{21}$	$\frac{20}{28}$	$\frac{25}{35}$	$\frac{30}{42}$	$\frac{35}{49}$	$\frac{40}{56}$	$\frac{45}{63}$	$\frac{50}{70}$	$\frac{55}{77}$	$\frac{60}{84}$	$\frac{65}{91}$	$\frac{70}{98}$	$\frac{75}{105}$	0.714	71.4%
$\frac{6}{7}$	$\frac{12}{14}$	$\frac{18}{21}$	$\frac{24}{28}$	$\frac{30}{35}$	$\frac{36}{42}$	$\frac{42}{49}$	$\frac{48}{56}$	$\frac{54}{63}$	$\frac{60}{70}$	$\frac{66}{77}$	$\frac{72}{84}$	$\frac{78}{91}$	$\frac{84}{98}$	$\frac{90}{105}$	0.857	85.7%
$\frac{1}{8}$	$\frac{2}{16}$	$\frac{3}{24}$	$\frac{4}{32}$	$\frac{5}{40}$	$\frac{6}{48}$	$\frac{7}{56}$	$\frac{8}{64}$	$\frac{9}{72}$	$\frac{10}{80}$	$\frac{11}{88}$	$\frac{12}{96}$	$\frac{13}{104}$	$\frac{14}{112}$	$\frac{15}{120}$	0.125	$12\frac{1}{2}\%$
$\frac{3}{8}$	$\frac{6}{16}$	$\frac{9}{24}$	$\frac{12}{32}$	$\frac{15}{40}$	$\frac{18}{48}$	$\frac{21}{56}$	$\frac{24}{64}$	$\frac{27}{72}$	$\frac{30}{80}$	$\frac{33}{88}$	$\frac{36}{96}$	$\frac{39}{104}$	$\frac{42}{112}$	$\frac{45}{120}$	0.375	$37\frac{1}{2}\%$
$\frac{5}{8}$	$\frac{10}{16}$	$\frac{15}{24}$	$\frac{20}{32}$	$\frac{25}{40}$	$\frac{30}{48}$	$\frac{35}{56}$	$\frac{40}{64}$	$\frac{45}{72}$	$\frac{50}{80}$	$\frac{55}{88}$	$\frac{60}{96}$	$\frac{65}{104}$	$\frac{70}{112}$	$\frac{75}{120}$	0.625	$62\frac{1}{2}\%$
$\frac{7}{8}$	$\frac{14}{16}$	$\frac{21}{24}$	$\frac{28}{32}$	$\frac{35}{40}$	$\frac{42}{48}$	$\frac{49}{56}$	$\frac{56}{64}$	$\frac{63}{72}$	$\frac{70}{80}$	$\frac{77}{88}$	$\frac{84}{96}$	$\frac{91}{104}$	$\frac{98}{112}$	$\frac{105}{120}$	0.875	$87\frac{1}{2}\%$
$\frac{1}{9}$	$\frac{2}{18}$	$\frac{3}{27}$	$\frac{4}{36}$	$\frac{5}{45}$	$\frac{6}{54}$	$\frac{7}{63}$	$\frac{8}{72}$	$\frac{9}{81}$	$\frac{10}{90}$	$\frac{11}{99}$	$\frac{12}{108}$	$\frac{13}{117}$	$\frac{14}{126}$	$\frac{15}{135}$	$0.\overline{1}$	$11\frac{1}{9}\%$
$\frac{2}{9}$	$\frac{4}{18}$	$\frac{6}{27}$	$\frac{8}{36}$	$\frac{10}{45}$	$\frac{12}{54}$	$\frac{14}{63}$	$\frac{16}{72}$	$\frac{18}{81}$	$\frac{20}{90}$	$\frac{22}{99}$	$\frac{24}{108}$	$\frac{26}{117}$	$\frac{28}{126}$	$\frac{30}{135}$	$0.\overline{2}$	$22\frac{2}{9}\%$
$\frac{4}{9}$	$\frac{8}{18}$	$\frac{12}{27}$	$\frac{16}{36}$	$\frac{20}{45}$	$\frac{24}{54}$	$\frac{28}{63}$	$\frac{32}{72}$	$\frac{36}{81}$	$\frac{40}{90}$	$\frac{44}{99}$	$\frac{48}{108}$	$\frac{52}{117}$	$\frac{56}{126}$	$\frac{60}{135}$	$0.\overline{4}$	$44\frac{4}{9}\%$
$\frac{5}{9}$	$\frac{10}{18}$	$\frac{15}{27}$	$\frac{20}{36}$	$\frac{25}{45}$	$\frac{30}{54}$	$\frac{35}{63}$	$\frac{40}{72}$	$\frac{45}{81}$	$\frac{50}{90}$	$\frac{55}{99}$	$\frac{60}{108}$	$\frac{65}{117}$	$\frac{70}{126}$	$\frac{75}{135}$	$0.\overline{5}$	$55\frac{5}{9}\%$
$\frac{7}{9}$	$\frac{14}{18}$	$\frac{21}{27}$	$\frac{28}{36}$	$\frac{35}{45}$	$\frac{42}{54}$	$\frac{49}{63}$	$\frac{56}{72}$	$\frac{63}{81}$	$\frac{70}{90}$	$\frac{77}{99}$	$\frac{84}{108}$	$\frac{91}{117}$	$\frac{98}{126}$	$\frac{105}{135}$	$0.\overline{7}$	$77\frac{7}{9}\%$
$\frac{8}{9}$	$\frac{16}{18}$	$\frac{24}{27}$	$\frac{32}{36}$	$\frac{40}{45}$	$\frac{48}{54}$	$\frac{56}{63}$	$\frac{64}{72}$	$\frac{72}{81}$	$\frac{80}{90}$	$\frac{88}{99}$	$\frac{96}{108}$	$\frac{104}{117}$	$\frac{112}{126}$	$\frac{120}{135}$	$0.\overline{8}$	$88\frac{8}{9}\%$

Note: The decimals for sevenths have been rounded to the nearest thousandth.

Reference

1–110 Grid

1	2	3	4	5	6	7	8	9	10
11	12	13	14	15	16	17	18	19	20
21	22	23	24	25	26	27	28	29	30
31	32	33	34	35	36	37	38	39	40
41	42	43	44	45	46	47	48	49	50
51	52	53	54	55	56	57	58	59	60
61	62	63	64	65	66	67	68	69	70
71	72	73	74	75	76	77	78	79	80
81	82	83	84	85	86	87	88	89	90
91	92	93	94	95	96	97	98	99	100
101	102	103	104	105	106	107	108	109	110

Multiplication/Divison Facts Table

$*, /$	1	2	3	4	5	6	7	8	9	10	11	12
1	1	2	3	4	5	6	7	8	9	10	11	12
2	2	4	6	8	10	12	14	16	18	20	22	24
3	3	6	9	12	15	18	21	24	27	30	33	36
4	4	8	12	16	20	24	28	32	36	40	44	48
5	5	10	15	20	25	30	35	40	45	50	55	60
6	6	12	18	24	30	36	42	48	54	60	66	72
7	7	14	21	28	35	42	49	56	63	70	77	84
8	8	16	24	32	40	48	56	64	72	80	88	96
9	9	18	27	36	45	54	63	72	81	90	99	108
10	10	20	30	40	50	60	70	80	90	100	110	120
11	11	22	33	44	55	66	77	88	99	110	121	132
12	12	24	36	48	60	72	84	96	108	120	132	144

Fraction-Stick and Decimal Number-Line Chart

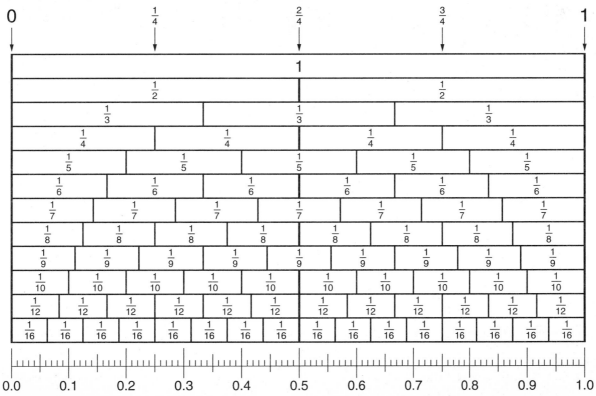

Reference

Metric System

Units of Length

1 kilometer (km)	= 1,000 meters (m)
1 meter	= 10 decimeters (dm)
	= 100 centimeters (cm)
	= 1,000 millimeters (mm)
1 decimeter	= 10 centimeters
1 centimeter	= 10 millimeters

Units of Area

1 square meter (m^2)	= 100 square decimeters (dm^2)
	= 10,000 square centimeters (cm^2)
1 square decimeter	= 100 square centimeters
1 are (a)	= 100 square meters
1 hectare (ha)	= 100 ares
1 square kilometer (km^2)	= 100 hectares

Units of Volume

1 cubic meter (m^3)	= 1,000 cubic decimeters (dm^3)
	= 1,000,000 cubic centimeters (cm^3)
1 cubic decimeter	= 1,000 cubic centimeters

Units of Capacity

1 kiloliter (kL)	= 1,000 liters (L)
1 liter	= 1,000 milliliters (mL)

Units of Mass

1 metric ton (t)	= 1,000 kilograms (kg)
1 kilogram	= 1,000 grams (g)
1 gram	= 1,000 milligrams (mg)

U.S. Customary System

Units of Length

1 mile (mi)	= 1,760 yards (yd)
	= 5,280 feet (ft)
1 yard	= 3 feet
	= 36 inches (in.)
1 foot	= 12 inches

Units of Area

1 square yard (yd^2)	= 9 square feet (ft^2)
	= 1,296 square inches ($in.^2$)
1 square foot	= 144 square inches
1 acre	= 43,560 square feet
1 square mile (mi^2)	= 640 acres

Units of Volume

1 cubic yard (yd^3)	= 27 cubic feet (ft^3)
1 cubic foot	= 1,728 cubic inches ($in.^3$)

Units of Capacity

1 gallon (gal)	= 4 quarts (qt)
1 quart	= 2 pints (pt)
1 pint	= 2 cups (c)
1 cup	= 8 fluid ounces (fl oz)
1 fluid ounce	= 2 tablespoons (tbs)
1 tablespoon	= 3 teaspoons (tsp)

Units of Weight

1 ton (T)	= 2,000 pounds (lb)
1 pound	= 16 ounces (oz)

System Equivalents

1 inch is about 2.5 cm (2.54)

1 kilometer is about 0.6 mile (0.621)

1 mile is about 1.6 kilometers (1.609)

1 meter is about 39 inches (39.37)

1 liter is about 1.1 quarts (1.057)

1 ounce is about 28 grams (28.350)

1 kilogram is about 2.2 pounds (2.205)

1 hectare is about 2.5 acres (2.47)

Rules for Order of Operations

1. Do operations within parentheses or other grouping symbols before doing anything else.

2. Calculate all powers.

3. Multiply or divide in order, from left to right.

4. Add or subtract in order, from left to right.

Units of Time

1 century	= 100 years
1 decade	= 10 years
1 year (yr)	= 12 months
	= 52 weeks (plus one or two days)
	= 365 days (366 days in a leap year)
1 month (mo)	= 28, 29, 30, or 31 days
1 week (wk)	= 7 days
1 day (d)	= 24 hours
1 hour (hr)	= 60 minutes
1 minute (min)	= 60 seconds (sec)

∗,/ Fact Triangles

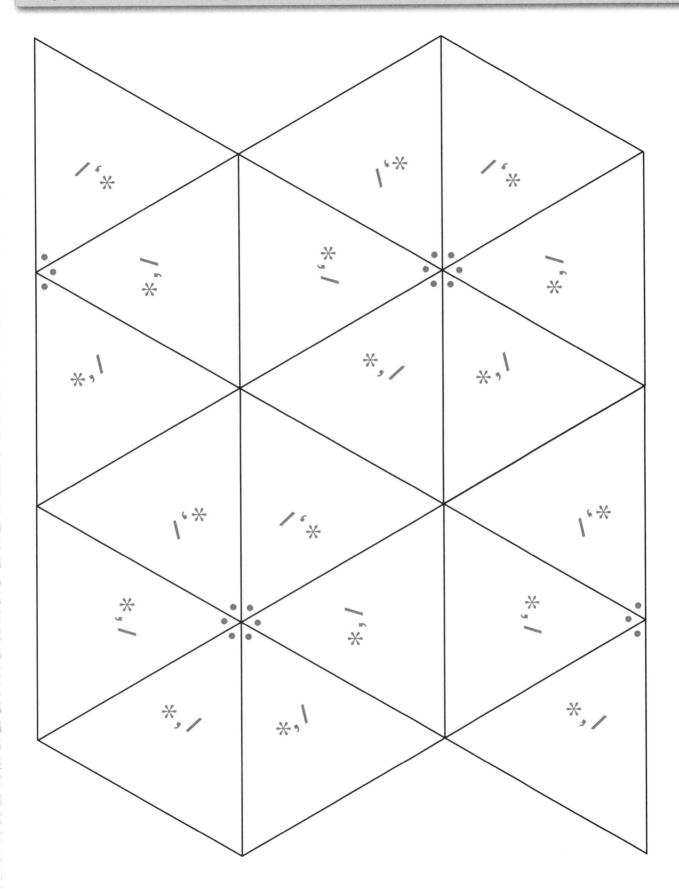

Grab-It Gauge

0.22 second _____

0.21 _____

0.20 _____

0.19 _____

0.18 _____

0.17 _____

0.16 _____

0.15 _____

0.14 _____

0.13 _____

0.12 _____

0.11 _____

0.10 _____

0.09 _____

0.08 _____

0.07 _____

0.00 **starting position**
     for contestant _____

Activity Sheet 2

Polygon Capture Pieces

Polygon Capture **Property Cards (Front)**

There is only one right angle.	There are one or more right angles.	All angles are right angles.	There are no right angles.
There is at least one acute angle.	At least one angle is more than 90°.	All angles are right angles.	There are no right angles.
All opposite sides are parallel.	Only one pair of sides is parallel.	There are no parallel sides.	All sides are the same length.
All opposite sides are parallel.	Some sides have the same length.	All opposite sides have the same length.	**Wild Card:** Pick your own side property.

Polygon Capture **Property Cards (Back)**

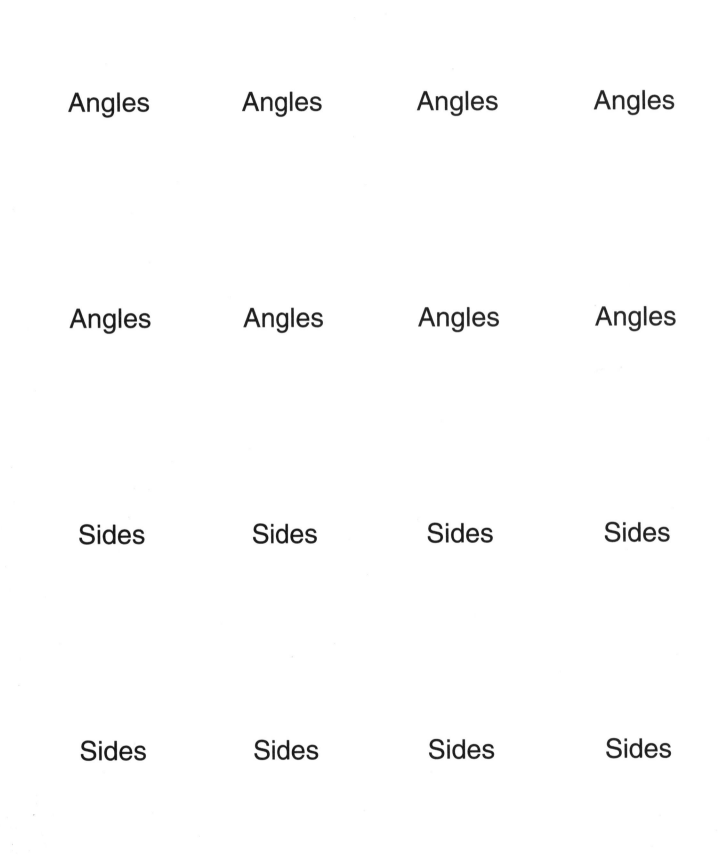

Angles   Angles   Angles   Angles

Angles   Angles   Angles   Angles

Sides   Sides   Sides   Sides

Sides   Sides   Sides   Sides